T0227412

THE NAMIB SAND SEA

The Namib Sand Sea

Dune forms, processes and sediments

By
N.LANCASTER
Department of Geology, Arizona State University

CRC Press
Taylor & Francis Group
Boca Raton London New York

CRC Press is an imprint of the
Taylor & Francis Group, an **informa** business

A BALKEMA BOOK

CIP-DATA KONINKLIJKE BIBLIOTHEEK, DEN HAAG

Lancaster, N.

The Namib Sand Sea: dune forms, processes and sediments / N.Lancaster. – Rotterdam [etc.]: Balkema. – Ill.
With ref.
SISO 567.3 UDC 551.311.31(688)
Subject heading: dunes; Namib Sand Sea.

Published by:
CRC Press/Balkema
P.O.Box 447, 2300 AK Leiden, The Netherlands
e-mail: Pub. NL@tandf.co uk
www.balkema.nl, www.tandf.co.uk, www.crcpress.com

© 1989 by Taylor & Francis Group, LLC
CRC Press/Balkema is an imprint of Taylor & Francis Group, an Informa business

No claim to original U.S. Government works

ISBN 13: 978-9-06191-697-0 (hbk)

Visit the Taylor & Francis Web site at
http://www.taylorandfrancis.com

and the CRC Press Web site at
http://www.crcpress.com

CONTENTS

ILLUSTRATIONS AND TABLES   VII

PREFACE   XI

1. INTRODUCTION   1
   1.1 Geological significance of aeolian sand accumulations   1
   1.2 Previous studies of sand seas   1
   1.3 Sand seas as depositional systems   2
   1.4 Aims and rationale of this study   6

2. REGIONAL PHYSIOGRAPHIC AND CLIMATIC SETTING   7
   2.1 Introduction   7
   2.2 Physiography of the Namib Desert   7
   2.3 Geologic and geomorphic history of the Namib   12
   2.4 Climate of the Namib   14
   2.5 Origin of the present climatic pattern   18

3. DUNE MORPHOLOGY AND MORPHOMETRY   20
   3.1 Crescentic dunes   20
   3.2 Linear dunes   28
   3.3 Star dunes   36
   3.4 Other dune types   41
   3.5 The pattern of dune size and spacing   42
   3.6 The pattern of dune alignments   45
   3.7 The pattern of dune morphology and morphometry in the sand sea 45

4. DUNE SEDIMENTS   47
   4.1 Sand colours   47
   4.2 Grain morphology   50
   4.3 Grain mineralogy   52
   4.4 Grain size and sorting characteristics of dune and interdune sands 52
   4.5 Spatial variation in grain size and sorting parameters   73
   4.6 Internal sedimentary structures of dunes   78

5. DUNE PROCESSES   81
    5.1 Introduction   81
    5.2 Winds and sand transport patterns   81
    5.3 Dune dynamics   94

6. CONTROLS OF DUNE MORPHOLOGY   111
    6.1 Factors influencing dune type   111
    6.2 Nature of the relationships between dune height and spacing   120
    6.3 Factors influencing dune size and spacing   123
    6.4 The development of aeolian bedforms   131

7. THE ACCUMULATION OF THE SAND SEA   132
    7.1 Models for sand sea accumulation   132
    7.2 Mechanisms for sand sea accumulation   134
    7.3 Development of the Namib Sand Sea   137
    7.4 The age of the Namib Sand Sea   146

REFERENCES   149

APPENDICES   161

INDEX   179

ILLUSTRATIONS AND TABLES

FIGURES

Figure 1. Location of major low latitude sand seas.
Figure 2. Southern Africa, showing location of Namib Desert on the west coast.
Figure 3. The Namib Desert: sand accumulations and major drainage systems.
Figure 4. The Namib Sand Sea: locality map.
Figure 5. Seasonal variation of wind directions in the central Namib Desert.
Figure 6. Daily cycle of wind direction in the central Namib Desert.
Figure 7. Distribution of different dune types in the Namib Sand Sea.
Figure 8. Morphometry of crescentic dunes.
Figure 9. Relationships between spacing of main and superimposed dunes.
Figure 10. Relationship between the height and spacing of crescentic dunes.
Figure 11. Surveyed profiles of complex and compound linear dunes.
Figure 12. Morphometry of compound and complex linear dunes.
Figure 13. Hierarchy of dune spacings in areas of compound and complex linear dunes.
Figure 14. Relationship between the height, width and spacing of linear dunes.
Figure 15. Surveyed profiles of star dunes.
Figure 16. Morphometry of star dunes.
Figure 17. Hierarchy of dune spacings in areas of star dunes.
Figure 18. Relationship between the height, width and spacing of star dunes.
Figure 19. Spatial variation of dune height and spacing in the Namib Sand Sea.
Figure 20. Pattern of dune alignments in the Namib Sand Sea.
Figure 21. Location of sites where dune sands were sampled.
Figure 22. Distribution of sand colours in the Namib Sand Sea.
Figure 23. Variation in grain roundness in the Namib Sand Sea.
Figure 24. Distribution of heavy mineral suites.
Figure 25. Dune types at sediment sampling locations.
Figure 26. Location of sampling points for dunes of different types.
Figure 27. Variation in grain size-frequency over representative crescentic dune.
Figure 28. Variation in grain size and sorting parameters over a representative crescentic dune.
Figure 29. Relationships between grain size and sorting parameters in areas of crescentic dunes.
Figure 30. Variation in grain size-frequency across a compound linear dune.
Figure 31. Variation in grain size-frequency across a complex linear dune.
Figure 32. Variation in grain size and sorting parameters across a representative complex linear dune.
Figure 33. Relationships between grain size and sorting parameters for compound linear dunes.

Figure 34. Relationships betwen the mean values of grain size and sorting parameters for complex linear dunes.

Figure 35. Variation in grain size-frequency across a star dune.

Figure 36. Variation in grain size and sorting parameters over a star dune.

Figure 37. Relationships between grain size and sorting parameters for star dunes.

Figure 38. Grain size-frequency and cumulative curves for zibar and sand sheets.

Figure 39. Relationships between grain size and sorting parameters for zibar and sand sheets.

Figure 40. Relationships between site mean values of grain size and sorting parameters for dunes of different types.

Figure 41. Relationships between mean grain size and sorting parameters for dunes of different types at site IX.

Figure 42. Truncation points on cumulative grain size-frequency curves for plinth and crest sands in areas of linear dunes.

Figure 43. Spatial variation in site mean values of grain size and sorting parameters for dune crest sands.

Figure 44. Size frequency histograms for crest and plinth sands in southern and northern parts of the sand sea.

Figure 45. Sedimentary structures in line.ir dunes.

Figure 46. Sedimentary structures near star dune crest.

Figure 47. Annual sand movement roses for sand sea area.

Figure 48. Seasonal variation in major sand transport directions in the northern parts of the sand sea.

Figure 49. Seasonal variation in resultant sand flow direction.

Figure 50. Relationship between magnitude and directional variability of sandflow.

Figure 51. Magnitude and frequency relationships between hours of sand moving winds and potential sand flow.

Figure 52. Patterns of wind velocity across crescentic dunes.

Figure 53. Relationships between mean speed up factors and dune shape and dune height for crescentic dunes.

Figure 54. Pattern of wind velocity over linear dunes and interdunes.

Figure 55. Relationship between mean speed up factor and linear dune shape.

Figure 56. Changes in surface wind direction across a complex linear dune indicated by wind ripple patterns.

Figure 57. Monthly pattern of linear dune activity.

Figure 58. Relationships between annual total of dune activity and total potential sand transport.

Figure 59. Spatial pattern of annual total dune activity across linear dunes.

Figure 60. Spatial patterns of annual net erosion and deposition across linear dunes.

Figure 61. Seasonal pattern of rate of advance of main slip face and east flank barchanoid dunes.

Figure 62. Ratio between plinth and crest sand transport rates for representative linear dune.

Figure 63. Ratio between plinth and crest sand transport for dunes of different heights at different wind velocities and directions.

Figure 64. Calculated pattern of erosion and deposition on linear dune for SSW and SW winds.

Figure 65. Patterns of erosion and deposition computed for cosine squared shape dune by Lai and Wu (1978).

Figure 66. Comparison of measured and predicted patterns of erosion and deposition on a complex linear dune during two week period of SSW and SW winds.

Figure 67. Comparison of measured and computed amounts of erosion and deposition on complex linear dune for two week period of SW and SSW winds.

Figure 68. Relationships between dune types and RDP/DP ratios.

Figure 69. Relationships between dune types, wind regimes and equivalent sand thickness for Namib Sand Sea.

Figure 70. The pattern of alignments in areas of complex linear dunes.
Figure 71. The pattern of dune alignments in areas of star dunes.
Figure 72. Relationships between dune types, wind regimes and equivalent sand thickness.
Figure 73. Relationships between the height and spacing of different dune types in different sand seas.
Figure 74. Relationship between bedform wavelength (spacing) and P_{20}.
Figure 75. Hierarchical arrangement of dune spacings in the Namib Sand Sea.
Figure 76. Relationships between dune spacing and P_{20} for the Namib Sand Sea.
Figure 77. Relationship between spacing of crescentic dunes and P_{20} in Namibian and Gran Desierto sand seas.
Figure 78. Simulated dune profiles at different wind velocities.
Figure 79. Relationship between dune height and total and resultant sandflow.
Figure 80. Relationship between equivalent sand thickness and total potential sandflow in the Namib Sand Sea.
Figure 81. Relationship between cross sectional area of dune and dune height for surveyed dunes.
Figure 82. Relationship between equivalent sand thickness and dune height for surveyed dunes.
Figure 83. Comparative relationships between grain size and sorting characteristics of fluvial, beach, shelf and dune sands from the southern Namib hinterland and coastal zone.
Figure 84. Relationships between coastal morphology and dune initiation in the Elizabeth Bay area.
Figure 85. Magnitude and direction of resultant potential sand transport in the Namib Sand Sea.
Figure 86. Spatial variation of equivalent sand thickness of bedforms in Namib Sand Sea.
Figure 87. Sites of exposures of interdune pond and marsh deposits in the northern part of the Namib Sand Sea.
Figure 88. Relationships between bathymetry and modern coastline in the Conception-Meob area.

PHOTOS

Photo 1. Landsat Image of the Namib Sand Sea.
Photo 2. Barchans and simple crescentic dunes southeast of Conception Bay.
Photo 3. Compound crescentic dunes south of Sylvia Hill.
Photo 4. Vertical aerial photograph of compound crescentic dunes with oblique linear ridges.
Photo 5. Compound linear dunes in the southern part of the sand sea.
Photo 6. Partly vegetated reticulate compound linear dunes in the eastern part of the sand sea.
Photo 7. Aerial view of complex linear dunes in the central part of the sand sea.
Photo 8. Ground view of western slope of large complex linear dunes.
Photo 9. Barchanoid dunes on east flank of a complex linear dune.
Photo 10. Simple linear dune on WSW-ENE alignment crossing corridor between complex linear dunes.
Photo 11. Chains of star dunes north of Tsondab Vlei.
Photo 12. Isolated star dune in southeastern part of the sand sea.
Photo 13. Partly vegetated sand sheet in southern part of sand sea.
Photo 14. Low rolling dunes without slipface development (giant zibar) south of the Uri Hauchab mountains.
Photo 15. Shrub coppice dunes on coast south of Walvis Bay.
Photo 16. Reversal of slip face orientation and crest of linear dune.

TABLES

Table 1. Relative areal extent of different dune types in the sand sea.

Table 2. Crescentic dune morphometry at sample sites.

Table 3. Morphometry of linear dunes at sample sites.

Table 4. Morphometry of star dunes at sample sites.

Table 5. Mean values of grain size and sorting parameters for crescentic and barchan dunes.

Table 6. Mean values of grain size and sorting parameters for linear dunes.

Table 7. Mean values of grain size and sorting parameters for star dunes.

Table 8. Mean values of grain size and sorting parameters for zibar and sand sheets.

Table 9. Grain size and sorting characteristics of dune crest sands from different dune types in different sand seas and dunefields.

Table 10. Summary of potential sandflow characteristics for DERU and KEP wind recorders.

Table 11. Mean speed-up ratios for crescentic dunes.

Table 12. Mean speed-up ratios for linear dunes.

Table 13. Comparative rates of barchan movement.

Table 14. Wind regime environments of different dune types in the sand sea.

Table 15. Exponents for power function relationship between dune height and spacing in different sand seas.

X

PREFACE

Most of the data on which this book is based was collected during the years 1979-1982, whilst I was a Research Associate with the Desert Ecological Research Unit (DERU) at the Namib Research Institute, Gobabeb, Namibia. I thank the C.S.I.R. and the Transvaal Museum for support and the Division of Nature Conservation, South West Africa, for facilities and permission to work in the Namib Park. Work was also carried out in areas under the control of Consolidated Diamond Mines (SWA) Pty. Ltd. who gave permission for entry to Diamond Area No. 1. During 1982, additional wind recorders were loaned by the South African Weather Bureau. Wind data was processed on a micro computer donated to DERU by the Wilfred Metje Foundation. John Ward, then of the Kuiseb Environmental Project (KEP), kindly allowed me access to unpublished wind data from recorders maintained by this project. During my stay at Gobabeb, I enjoyed field support and stimulating discussions with Mary Seely (the Director of DERU), John Ward, Liz McLain and Ian Livingstone.

The book was largely written whilst I was on the staff of the University of Cape Town, where I received encouragement and support from Professor Richard Fuggle of the Department of Environmental and Geographical Science. The publication of this book was made possible by a generous grant from the Editorial Committee of the University of Cape Town. The aerial photographs were reproduced under South African Government Printer's copyright authority 7843 of 13.8.82.

Sections of this book have been reviewed by John Rogers, John Ward, Haim Tsoar, Gary Kocurek, Robert Folk, Ronald Greeley and the late E.D. McKee. I am very grateful for their comments and suggestions for its improvement.

Neither the work on which this book is based, nor its writing, would have been possible without the assistance of my wife, Judith, who has at various times acted as field assistant, typist, draftsperson, laboratory assistant and computer operator.

1 INTRODUCTION

1.1 GEOLOGICAL SIGNIFICANCE OF AEOLIAN SAND ACCUMULATIONS

Desert sand seas, or ergs, contain more than 95% of the world's aeolian sand. Of this, 85% is contained in sand seas with an area exceeding 32,000 km^2 (Wilson 1973). Major areas of sand sea accumulation lie in the old world deserts of the Sahara, Arabia, central Asia, Australia and southern Africa, where sand seas cover between 20 and 45% of the area classified as arid (Fig. 1). In North and South America there are no large active sand seas and dunes cover less than 1% of the arid zone. There is evidence, in the form of dune systems fixed by vegetation, that sand seas in subtropical latitudes were much more extensive in the period coeval with the last Glacial maximum, some 18,000 years BP (Sarnthein 1978; Goudie 1983); whilst many north American dunefields were more extensive in the mid Holocene (Smith 1965; Ahlbrandt et al. 1983). There are large aeolian sand accumulations on other planets and the North Polar Sand Sea of Mars is the largest known sand sea on the terrestrial planets, with an area of 7-8 x 10^5 km^2 (Tsoar et al. 1979).

In the rock record, sandstones considered to be largely aeolian in origin have been identified from the Precambrian (e.g. Ross 1983) to the Tertiary (e.g. Ward 1984). Their characteristics are discussed by McKee and Bigarella (1979) and Mader and Yardley (1985). Aeolian sandstones are particularly common in formations of Permian to Triassic age (McKee and Bigarella 1979; Mader 1983) when they occurred on most continents. The majority of aeolian sandstones apparently accumulated in low latitude desert regions as sand bodies which are comparable in extent and environmental setting with modern sand seas (Glennie 1970; Kocurek 1981a, b; Lupe and Ahlbrandt 1979). The best way to understand the origins and depositional environments of aeolian sandstones is to study modern aeolian deposits, as demonstrated by Glennie (1970), Hunter (1977), McKee (1979) and Ahlbrandt and Fryberger (1982).

1.2 PREVIOUS STUDIES OF SAND SEAS

Despite their environmental and sedimentary importance, few modern active sand seas have been the subject of detailed study. In part this has been a logistic problem, as many sand seas are located in regions which even today are remote and difficult of access.

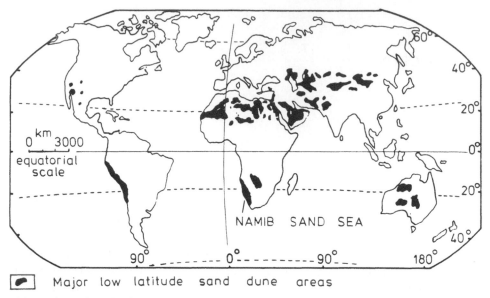

Figure 1. Location of major low latitude sand seas.

Environmental information, especially in the form of climatic records, is unavailable for many desert areas. This often prevents correlation of dune forms with wind patterns and the estimation of sand transport rates and directions.

Implicit in many studies of sand seas is an inductive, empirical and historical approach, which has focussed on the development of the sand sea through geologic time. The studies of Capot-Rey (1947), Alimen et al. (1958), and Chavaillon (1964) in the Sahara; Allchin, Goudie and Hegde (1978) and Wasson et al. (1983) in India; Whitney et al. (1983) in Saudi Arabia; Wopfner and Twidale (1967), Wasson (1983 a,b) in Australia; Ahlbrandt and Fryberger (1980) in the USA and Besler (1976, 1980) in the Namib fall into this category.

The availability of remote sensing imagery of desert sand seas, especially the satellite imagery provided by Landsat in the 1970's and 1980's has promoted studies of the patterns of dunes in sand seas (e.g. McKee and Breed 1976; Mainguet and Callot 1978; Breed et al. 1979) and drawn attention to the relationships between dunes in different parts of the same sand sea and between the sand sea and its surrounding area (Breed et al. 1979). In particular, there has been the realisation that dunes of essentially similar form occur in widely separated sand seas, both on Earth and Mars (Breed 1977; Breed and Grow 1979). There has also been the recognition that many sand seas form part of regional and sub continental scale sediment transport systems (Mainguet 1977, 1983, 1984; Wilson 1971; Fryberger and Ahlbrandt 1979).

1.3 SAND SEAS AS DEPOSITIONAL SYSTEMS

Sand seas constitute important depositional systems, or groups of process related sedimentary facies, both today and throughout much of geologic time. The sand sea

2

depositional system contains sources and sinks for sediment, linked by a cascade of energy and materials which can be viewed in terms of sediment inputs and outputs, transfers and storages in a manner analogous to the drainage basin in fluvial geomorphology (Cooke and Warren 1973; Chorley and Kennedy 1971). The systems appoach is well established in geomorphic studies and focusses attention on the relationships between processes and materials which produce a characteristic set of landforms or responses. Implicit in this approach are concepts of dynamic equilibrium between processes and forms. Such concepts may not always be easily transferred to studies of depositional landscapes, such as alluvial fans and desert sand seas (Bull 1975), because time is not an independent variable in their formation. In addition, the temporal and spatial scales involved are often large (Schumm and Lichty 1965). However, the systems approach does make it possible to recognise a hierarchy of system components and to analyse the operation of processes at different spatial and temporal scales. Outlined below is a general model for a sand sea system, which will serve as a conceptual framework for this book.

The principal energy source for the sand sea system is the kinetic energy of the wind, derived from regional and local scale atmospheric circulations driven by variations in the amount of solar energy received at the earth's surface.

1.3.1 *Inputs*

The accumulation of a desert sand sea requires that there is a source of sand sized sediment external to the sand body. Most sand seas are composed of quartz sand, which is derived ultimately from the weathering of quartz rich rocks, especially granites and sandstones. Direct contributions of sand to sand seas from the weathering of bedrock are probably limited, although sandstones, some of which are aeolian, have been cited as sources of sand for dunefields in north America (Ahlbrandt 1974) and elsewhere (Besler and Marker 1979; Wasson et al. 1983). Most aeolian sand is derived from material which has been transported by some other medium. It is mobilised by deflation, so that most sand seas lie downwind of areas with high deflation rates (Wilson 1971). In mixed sediments, deflation ceases once a protective armour or lag of coarse particles is developed, so most sources of aeolian sand are also areas of active sediment renewal. As an example, Sharp (1980) noted that sand transport rates in the Coachella Valley, California, as reflected in abrasion of test objects, increased markedly after flooding of the source of sand in the Whitewater Wash.

There are a variety of sources of aeolian sediment for sand seas. The most important are probably fluvial and deltaic sediments. In the Sahara, most earlier workers stated that the ergs were derived from fluvial sediments deposited in earlier and more humid periods (e.g. Alimen et al. 1958; Chavaillon 1964; Capot Rey 1970; Dresch 1982). Fluvial and alluvial fan sediments are important sources for north American dunefields (Beheiry 1967; Sharp 1966; Merriam 1969; Andrews 1981), as well as for sand seas in Arabia (Glennie 1970; Besler 1982); India (Wasson et al. 1983b) and Australia (Wasson 1983). Playas and sebkhas may constitute a source of sediment for dunefields such as White Sands, New Mexico (McKee 1966), some areas of Oman (Besler 1982) and the Simpson-Strzelecki dunefield of Australia (Wasson 1983). Many cold climate dunefields were fed by deflation of glacial outwash areas (Smith 1965; David 1977; Carter 1981).

3

Beaches, both lacustrine and marine, are sources of sand for dunefields. The shorelines of former Lake Cahuilla appear to have been the source for the Algodones dunes in California (Norris and Norris 1961; McCoy et al. 1967). Marine beaches provide sediment sources for dunes in Baja California (Inman et al. 1966); India (Goudie and Sperling 1977); Peru (Segestrom 1962) and the Namib (Rogers 1977, Lancaster 1982a). Some sand seas, such as those in Australia and the Kalahari, have internal sediment sources as a result of the deflation of interdunes and reactivation of older dunes (King 1960; Folk 1971; Wasson 1983b). Many sand seas receive sediment from a number of sources and in practice it is often difficult to precisely identify source areas, especially as they may be located at some distance from the sand sea (Fryberger and Ahlbrandt 1979). Further, in the Sahara and Arabia, many sand seas receive inputs of sand from adjacent sand bodies (Fryberger and Ahlbrandt 1979; Mainguet 1984).

1.3.2 *Transfers*

The main mechanism for the transfer of sediment from source areas to sand seas is transport of sand by the wind. Areas of light tones on satellite imagery often identify the position of such corridors for sand transport, as demonstrated by Mainguet (1984) in the Sahara. Features such as sand choked river valleys, sand sheets and streaks, barchans, shadow dunes and shrub coppice dunes are found along sand transport paths in many deserts (Breed et al. 1979; Fryberger and Ahlbrandt 1979; Lancaster 1982a; Mainguet 1977, 1984). Very little is known of the magnitude of sand transport along these corridors. Both transport by saltation (bed load transport) and by dunes (bedform transport) occur, with saltation apparently dominating. In the western Sahara, Sarnthein and Walger (1974) estimated that 62.5-162.5 $m^3.m^{-1}.yr^{-1}$ of sand was being transported in saltation, compared to 1.16-3.49 $m^3.m^{-1}.yr^{-1}$ as bulk transport by barchans. In Peru, Finkel (1959) calculated a total sand transport rate of 39.9 $m^3.m^{-1}.yr^{-1}$. In the same area, the bulk transport rate by barchans was estimated to be 0.4-0.743 $m^3.m^{-1}.yr^{-1}$ by Lettau and Lettau (1969). Rates of sand transport on sand sheets in the Jafurah area, Saudi Arabia measured by Fryberger et al. (1984) ranged between 2.3 and 21.5 $m^3.m^{-1}.yr^{-1}$, with a mean of 10.2 $m^3.m^{-1}.yr^{-1}$. Within the sand sea itself, sand is transferred towards sediment sinks both by dune migration and saltation of sand from dune to dune. The latter process is particularly important in producing the differentiation of grain size and sorting parameters at dune and sand sea scales (Lancaster 1982b).

1.3.3 *Storages*

Two types of sediment storage take place in the sand sea system. The first is storage along transport paths as sand drifts, shadows and streaks, as well as in small dunes. Much of this storage is probably short term compared to sediment storage in the dunes of the sand sea itself. The work of Wilson (1971), Mainguet and Chemin (1983) and Mainguet (1984) has shown that sand seas are not merely sinks for aeolian sand. They are dynamic sedimentary bodies which gain and loose sand. The dunes of the sand sea therefore represent long term sediment storage, in a manner which is analogous to sediment storage in river flood plains. Sand is stored in the sand sea in three main ways: dune storage, mega

4

dune or draa storage and interdune storage. Sand is transferred to storage, i.e. deposited, by bedform climbing, as well as by grainfall and grainflow deposition on lee side avalanche faces. Where interdunes are damp or wet, deposition by adhesion ripples may also occur. Deposits of these types are the primary units of aeolian deposition (Hunter 1977). Growth of large bedforms, termed mega dunes, compound and complex dunes (McKee 1979) or draa (Wilson 1971) occurs by the merging or modification of smaller dunes (Mader and Yardley 1985) and by vertical growth in opposed or complex wind regimes (Lancaster 1983a). Both modification and merging involve a degree of bedform climbing, with small superimposed dunes migrating over larger forms to create second order bounding surfaces (Brookfield 1977). Migration of large or small dunes, with bedform climbing as a result of downwind decreases in sediment transport rates (Rubin and Hunter 1982), transfers sediment to the sediment sink and leads to the accumulation of the sand body as a whole.

The temporal and spatial scales involved in the different levels of storages may be expressed in terms of the bedform reconstitution time of Allen (1974). This is the time that is required for the bedform to migrate one whole wavelength downcurrent and increases with the size of the bedform. For small dunes, it is a matter of years. Large dunes or draa may take 10^2 to 10^4 years to move one wavelength, whilst for sand seas reconstitution time is measured on time scales of 10^6 to 10^8 years.

1.3.4 *Outputs*

Although Wilson (1971), Mainguet (1977) and Mainguet and Chemin (1983) have shown that the same winds which transfer sand to the sand sea can also remove it, loss of sand from the downwind margins of many sand seas and dunefields is often minimal, especially if they are composed of migrating 'sand trapping' bedforms, such as crescentic dunes. In this case the margin of the sand sea will occur at the point which the migrating dunes have reached downwind of their original source. Where the bulk of the dunes in the sand sea are of the 'sand passing' variety, the downwind margin will shed sand if the dunes of the sand sea are fully developed to equilibrium with the overall sand transport conditions. As Wilson (1971) has noted, this is rarely achieved, because of the considerable time required for dunes to achieve full equilibrium with sand flow conditions and the inconstancy of these conditions due to climatic changes. Alternatively, the downwind margin of the sand sea may be determined by an increase in sand transport rates as a result of regional climatic changes, such that sand removal exceeds the rate of replacement from upwind. Mainguet and Chemin (1983) have suggested that the margins of sand seas like the Erg Chech are determined in this way.

Loss of sand to fluvial or marine processes occurs in some sand seas. Dunes on the Gulf coast of Arabia and in the western Sahara have been shown to be prograding into shallow seas (Shinn 1973; Sarnthein and Diester-Haas 1977; Fryberger et al. 1983). Barchans and transverse dunes prograde into sebkhas and salt marshes on the northern margins of the Namib Sand Sea (Nagtegaal 1973; McKee 1982) and dunes are being actively eroded by wave action on its western margin between Sylvia Hill and Sandwich Harbour (Rogers 1979). Fluvial erosion of sand sea margins occurs along the Kuiseb river in the Namib (Ward and Von Brunn 1985); in Oman (Glennie 1970); along the Wadi

Saoura in the northern Sahara (Mabbutt 1977) and on the margins of the Great Sand Dunes, Colorado (Andrews 1981).

1.4 AIMS AND RATIONALE OF THIS STUDY

The work of Wilson (1971), Mainguet (1977, 1984), Mainguet and Chemin (1983) and Fryberger and Ahlbrandt (1979) has shown clearly that modern sand seas are dynamic sedimentary bodies with inputs and outputs of aeolian sand. Given adequate environmental data, especially on wind velocity and direction, it is possible to study sand seas in a dynamic systems framework. The Namib Sand Sea is ideally suited to such an approach. It is relatively small (34,000 km^2) and environmental information is freely available, especially for the northern parts of the area. A wide variety of dune types occurs in the sand sea, facilitating examination of the factors influencing dune morphology.

The aim of this book is to show that the Namib Sand Sea is a dynamic depositional system and that the relationships between dune forms, processes and sediments reflect this. The external morphology and surface sediments of the dunes contain information which shows how the sand sea has accumulated. This can be used to develop models of dune morphology and sand sea accumulation.

The Namib Sand Sea is first considered in its regional physiographic and climatic setting (Chapter 2). Chapters 3 and 4 examine the morphology and morphometry of the dunes and the characteristics of their sediments. In Chapter 5, the aeolian processes operating at the sand sea and dune scales are discussed. Having considered the main components of the sand sea system and their operation, Chapters 6 and 7 are integrative and analyse the factors that control dune morphology and that are responsible for the accumulation of the sand sea.

2 REGIONAL PHYSIOGRAPHIC AND CLIMATIC SETTING

2.1 INTRODUCTION

The Namib Desert extends for over 2000 km along the west coast of southern Africa from the Olifants River in South Africa (32°S) to the Carunjamba River at 14°S in Angola (Fig. 2). Inland, it is bounded by the Great Escarpment, which lies 120-200 km from the coast and forms the western edge of the interior plateau of southern Africa (Wellington 1955).

The Namib is one of five west coast subtropical deserts in the world, the others being in western Australia, northwest Africa, Baja California and Chile-Peru (Meigs 1966). With the exception of western Australia, they are all hyper-arid as a result of the upwelling of cold waters offshore. Both the Atacama and Baja California deserts lie in areas of active tectonism and are restricted to narrow coastal plains. Northwestern Africa shares many of the tectonic and geomorphic characteristics of the Namib, but constitutes a western extension of the interior Sahara desert. The Namib is unique amongst the west coast deserts in that it lies in an area of relative tectonic stability, yet is crossed in its southern parts by a major perennial river, the Orange, which carries a major part of the sediment eroded from the interior of southern Africa (Rogers 1977; Dingle and Hendey 1984). This is of vital importance to the origins and development of the Namib Sand Sea, which lies in the central part of the desert, between latitudes 26 and 23°S, downwind of the mouth of the Orange River at 28°S (Fig. 3). Other important sand seas and dunefields in the Namib include the Obib dunes on the north bank of the Orange River; the Skeleton Coast dunefield (Lancaster 1982a), the Cunene Sand Sea (Bremner 1984) and the Curoca-Bahia dos Tigres dunefield in Angola (Torquato 1972). All these sand formations lie to the north (downwind) of important rivers which flow either perennially (the Orange and Cunene) or seasonally (the Ugab, Uniab and Hoanib).

2.2 PHYSIOGRAPHY OF THE NAMIB DESERT

The Namib can be subdivided into four main areas: the southern or transitional Namib, which includes coastal Namaqualand and the Sperrgebiet; the Namib Sand Sea; the central Namib Plains and the northern Namib and Skeleton Coast.

South of the Orange River lies the Namaqualand Sandy Namib, which recieves a

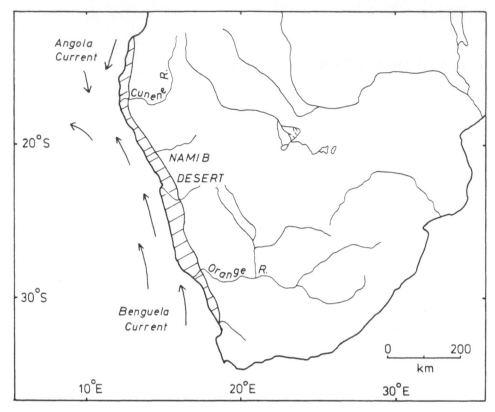

Figure 2. Southern Africa showing location of Namib Desert on the west coast.

winter rainfall of 50-150 mm that is sufficient to maintain a succulent vegetation cover. This vegetation stabilises the surficial sands, most of which were brought to the coast by local rivers during the last Glacial period and redistributed as small dunefields (Rogers 1977; Tankard and Rogers 1978).

North of the Orange River, between Oranjemund and the Koichab valley is the Sperrgebiet, which consists of extensive rocky and sand covered plains extending coastwards from the escarpment, with low hills at intervals. The Sperrgebiet plain is considered by Ollier (1977) and SACS (1980) to be continuous with similar surfaces to the north of the sand sea (the Namib Platform) and to form the basal unconformity below Cenozoic deposits in the region. South of Luderitz, there are sandy plains and small dunefields, with the sands derived directly from the Orange River. The Trough Namib (Wannennamib of Kaiser 1926) is a distinctive zone of wind eroded bedrock and yardang fields between Chamais Bay and the Grillental.

The Namib Sand Sea has an area of some 34,000 km² (Barnard 1973) and lies between Luderitz and the Kuiseb River. It extends inland to the base of the Great Escarpment at around the 1000 m contour (Fig. 4). A further small area of dunes extends north to the Swakop River along the coast, whilst to the south, trains of barchans and crescentic dunes extend north to the sand sea from the beaches of Elizabeth and Chamais Bays.

8

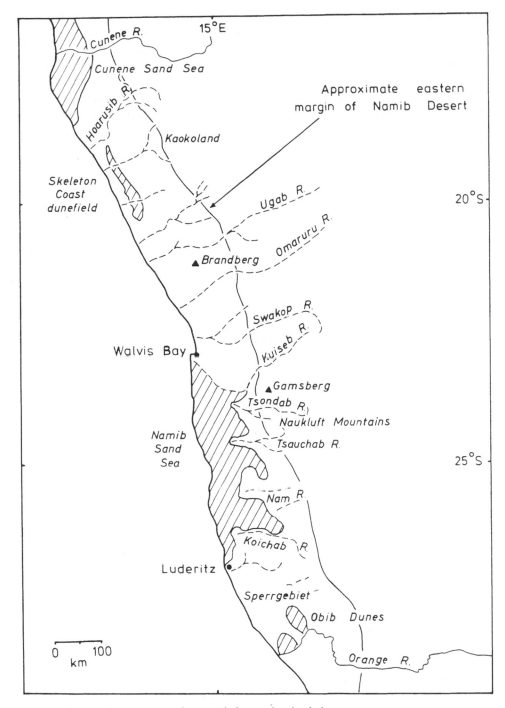

Figure 3. The Namib Desert: sand accumulations and major drainage systems.

In common with many sand seas, the margins of the Namib Sand Sea are well defined. To the north, except in its delta region, the Kuiseb River forms a distinct boundary, a role which it has probably exercised for much of the late Cenozoic (Ward 1984). The tips of linear dunes fall into a canyon like valley which is 100-200 m deep east of Gobabeb. Downstream of this point, the valley is shallow and broad. Its position and north westerly orientation appear to be controlled in part by the encroachment of linear and crescentic dunes from the south and the course of the stream is deflected around the tips of the dunes (Ward and Von Brunn 1985). There is evidence to suggest that it has been displaced by up to 30 km northward since the early-mid Pleistocene (Ward 1984). The duration and frequency of flows in the Kuiseb decreases downstream. Whilst the river flowed annually past Gobabeb during the period 1963-1979 (Seely et al. 1980) it only reached the Atlantic Ocean once during this period, and has been known to flow to the coast only fifteen times since 1837 (Stengel 1964).

The eastern margin of the sand sea lies at the base of the dissected western escarpment of the southern African plateau, which rises from an average altitude of 1000 m at its base to 1500-2500 m along its highest points in the Gamsberg and the Naukluft Mountains. The escarpment is essentially an erosion feature cut in Damara Sequence mica schists to the north and Namaqua Province gneisses and granites to the south (Selby 1977). Capping the escarpment south of the Naukluft Mountains are a series of early Paleozoic (550 Ma) intracratonic shallow marine and terrestrial sediments. Shales, sandstones and dolomitic limestones of the Nama Group provide a distinctive clastic input of oblate black and pale blue cobbles and pebbles to the Tsondab and Tsauchab Rivers. A remnant of the Jurassic Etjo Formation Sandstone of the Karoo Sequence caps the Gamsberg on the escarpment edge.

There is a sharp contrast in surface character and elevation between the dunes of the sand sea and the calcified Cenozoic alluvial fans which extend west from the escarpment. In places the dunes lap up against outliers of the escarpment. Between the valleys of the Tsauchab and Tsondab Rivers the dunes of the sand sea lie on a 60-90 m exposed thickness of Tertiary semi-consolidated sandstones of mainly aeolian origin (the Tsondab Sandstone Formation).

Within the sand sea, sand cover is continuous south of the Tsondab Valley. Inliers of Precambrian rocks are mostly confined to the southern and eastern parts of the sand sea, where inselbergs as well as larger mountain masses such as the Uri Hauchab and Guinasib Mountains introduce extradune alluvial and talus deposits into the sand sea and induce diversion of winds and dunes around them. North of the Tsondab Valley, sand cover is discontinuous and relict fluvial gravels of the Kuiseb and Tsondab Rivers are exposed in interdune areas. Small areas of interdune pond and pan deposits are locally preserved in interdunes between the Kuiseb and Tsondab Rivers (Teller and Lancaster 1986a).

A number of ephemeral watercourses drain towards the coast from the escarpment (Figs 3 and 4). The Kuiseb and rivers to the north have courses which extend to the coast and have been observed to flow to the sea at intervals (e.g. 1933, 1985). Rivers such as the Tsondab and Tsauchab, although they flow ephemerally today (Stengel 1970), have well defined valleys which reach depths of 80 and 150-200 m respectively and are incised into the Tsondab Sandstone Formation. They penetrate the sand sea for distances of 40-80 km from its eastern margins and terminate amongst the dunes in extensive playas. West of the

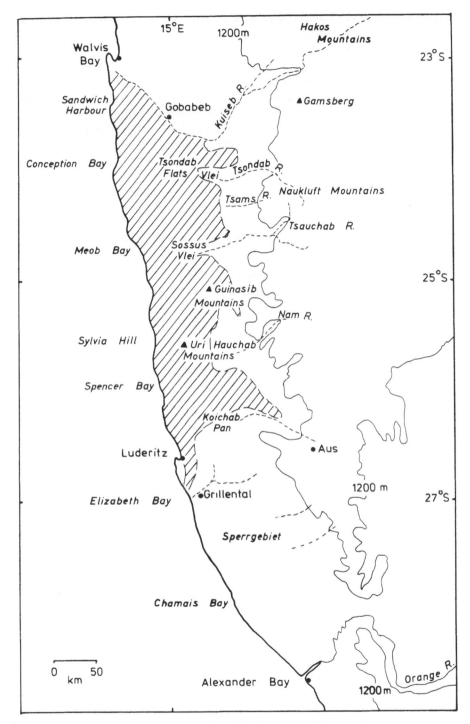

Figure 4. The Namib Sand Sea: locality map. Sand sea is shaded.

11

present end points of both the Tsondab and Tsauchab Rivers there are outcrops of fluvial deposits indicating that these rivers formerly flowed to points further to the west (Seely and Sandelowsky 1974; Lancaster 1984a). Very small ephemeral streams such as the Nam and Tsams terminate in small playas against the eastern margin of the sand sea.

The sand sea abuts the Atlantic coast of south-western Africa. Between Sylvia Hill and Meob Bay and again between Conception Bay and Sandwich Harbour, the dunes run into the sea and are eroded by wave action (Rogers 1979). Powerful longshore transport of sand towards the north gives rise to large coastal spits at Sandwich Harbour and Walvis Bay, which have accreted significantly in historical times (Logan 1960). The spits enclose extensive lagoons and coastal sebkhas (Nagtegaal 1973; McKee 1982). In the Conception-Meob area and at Hottentots Bay, there are extensive bedrock outcrops and coastal sebkhas where the Namib Platform comes to the surface. At Sylvia Hill and Spencer Bay, outcrops of late Precambrian schists and marbles form low hills along the shore.

North of the sand sea, between the Kuiseb River and the Brandberg is an extensive rocky plain, termed the Namib Platform by Logan (1960) and Ollier (1977), with an average gradient of only 1° between the coast and the 1000 m contour (Hovermann 1978). The plain is cut in mica schists and syntectonic granitic intrusions of the late Proterozoic geosynclinal Damara Sequence. Rising from the plain are low ranges of hills and isolated inselbergs (Ollier 1977; Selby 1977). Extensive pedogenic and groundwater gypsum and calcrete horizons are developed where relief is low (Goudie 1972; Watson 1979).

The dissected pediplain and sandstone and lava hills of the northern Namib form the area known as the Kaokoveld. The coastal area, which is partly covered by small dunefields, is known as the Skeleton Coast.

2.3 GEOLOGIC AND GEOMORPHIC HISTORY OF THE NAMIB

The geological history of the Namib as a distinct physiographic entity begins with the breakup of West Gondwana in the late Mesozoic (135-120 Ma) and the opening of the South Atlantic ocean. Subsequent events have been influenced by the position of the Namib on the trailing margin of the African crustal plate, periodic epeirogenic uplift of the continental margins along the escarpment hinge zone and world-wide eustatic changes in sea level.

The formation of the Namib began in the late Cretaceous, following the breakup of West Gondwana and the opening of the South Atlantic by tensional rifting which began some 135 Ma ago (Tankard et al. 1982; Dingle et al. 1983). By 80 Ma, fully marine conditions existed in the proto South Atlantic as the Falkland Plateau had cleared the tip of southern Africa at 100 Ma. The Great Escarpment was formed 127 Ma ago by tilting of the continental margin (Dingle and Scrutton 1974) and the Namib Platform was then cut by scarp retreat and pediplanation across the schists and granites of the Damara Sequence which formed the continental margin (Ollier 1977; Selby 1977). Sediments eroded from the escarpment zone, augmented by those from the Orange River were deposited as sediment wedges 4 km thick in the Orange Basin and 3 km thick in the Walvis Basin (Dingle et al. 1983). Sedimentation rates declined substantially after the late Cretaceous, suggesting stabilisation of the continental margin and a decline in rainfall in the interior of the continent (Dingle et al. 1983).

Although early Tertiary deposits in the northern Namib indicate mostly marine conditions in the early-middle Tertiary, sedimentation in the southern and central areas was exclusively terrestrial following the Eocene transgression. Widespread accumulation of terrestrial red brown sandstones, notably the Tsondab Sandstone Formation, occurred throughout the region during the mid to late Tertiary in conditions of arid to semi- arid climates.

The Tsondab Sandstone Formation is generally 45-90 m thick, but reaches a thickness of over 200 m in the eastern part of the Namib Sand Sea (Besler and Marker 1979). It is composed of subangular to rounded, well to poorly sorted, fine to medium quartz sands with an orange to red brown iron patina and is poorly consolidated, with a patchy carbonate, dolomitic and locally gypsum cement. Internal structures exposed at Tsondab Vlei and in the west of the sand sea show large scale cross bedding with dips towards the north and north-east, indicative of large aeolian dunes formed under a southerly wind regime (Ward et al. 1983). Near the eastern margins of the present Namib Sand Sea, fluvial and playa facies of the Tsondab Sandstone Formation occur (Ward 1984), implying fluvial input from the escarpment zone. Extensive horizons of pedotubules occur at intervals in the sandstone, suggesting subaerial exposure to termite activity and vegetation growth. The Tsondab Sandstone Formation and its equivalents in the southern and northern Namib represent a precursor of the present Namib Sand Sea, and were apparently deposited under a very similar wind regime to that of today in a period during the Palaeogene, perhaps coeval with the extensive Oligocene regression, which lowered sea level by up to 500 m (Seisser and Dingle 1981) and shifted the coastline by 100- 200 km westwards. This event possibly provided a source of sand for the Tsondab Sandstone Formation dunes (Ward 1984).

In the area of the Kuiseb Valley, the Tsondab Sandstone Formation is overlain unconformably by the well rounded quartzite and vein quartz gravels of the Karpfenkliff Conglomerate Formation (Ward 1984). The Karpfenkliff Conglomerate and its equivalents in the Tsondab and Tsauchab valleys represent the earliest evidence of a well integrated drainage system in the central Namib. They have been tentatively assigned a middle Miocene age by Ward et al. (1983) on the basis of correlations with lithologically similar gravels in the southern Namib (the Arriesdrift Gravel Formation) which contain a rich middle Miocene (10-18 Ma) fauna suggestive of a more mesic climate with savanna grasslands and gallery forest in this part of the Namib (Corvinus and Hendey 1978).

In the southern Namib, fossiliferous clays near Luderitz and in the Grillental (the Grillental Beds of the Elizabeth Bay Formation) contain a rich Lower Miocene mammalian fauna which is indicative of widespread wet conditions with shallow lakes. They are overlain by thin alluvial fan deposits and suceeded by the northward dipping cross-beds of the aeolian Fiskus Beds of the same formation. In this part of the Namib, these sandstones represent the first unequivocal indication of arid climates and are assigned a late Miocene to Pliocene age (SACS 1980).

Following their deposition, Tertiary fluvial deposits in the central Namib were extensively calcified. Concurrently a pedogenic calcrete up to 5 m thick developed in the Tsondab Sandstone Formation on adjacent interfluves. In the eastern central Namib, Yaalon and Ward (1982) suggested that formation of this horizon represented at least

13

500,000 years of landform stability in a semi-arid climate. The calcrete is correlated with similar deposits in the southern Namib of probable end Miocene age (SACS 1980; Ward et al. 1983) and provides an important regional stratigraphic marker.

During the late Tertiary, all major rivers in the Namib underwent extensive incision, as a result of epeirogenic uplift of the southern African subcontinent by 500 m. Quaternary fluvial deposits are confined to the valleys cut during this period (Gevers 1936; Mabbutt 1952; Korn and Martin 1957) and provide a record of periods of incision and aggradation of probable regional extent. The sequence in the Kuiseb is best known, having been studied by many workers (e.g. Rust and Wienecke 1974, 1980; Marker 1977; Ward 1982, 1984) and presently provides the type fluvial sequence for the region.

During the early to mid Pleistocene, quartzose gravels were deposited within the Kuiseb Valley by a braided stream with a greater competence than the modern Kuiseb, probably in an arid or semi arid environment (Ward 1984). Interbedded wedges of crossbedded quartz arenite up to 14 m thick are interpreted by Ward (1982, 1984) as paleo-dunes, possibly of linear type. Throughflow in a fluctuating groundwater body resulted in the cementation of the gravels by carbonate (Yaalon and Ward 1982), to form the the Oswater Conglomerate prior to re-excavation of the Kuiseb Valley and an incision of up to 50 m in the canyon reach during the mid-late Pleistocene (Ward 1984).

The 30 m thickness of the Homeb Silt Formation was deposited during the period 23,000-19,000 BP (Vogel 1982) as the flood-plain of the Kuiseb River aggraded (Ward 1984) rather than being deposited behind a dune dam, as suggested by Rust and Wienecke (1974,1980), or at the terminal playa of the river (Marker and Muller 1978; Vogel 1982).

A terminal Pleistocene re-incision phase removed much of the Homeb Silt Formation deposits from the main valley and also deposited the pebble to cobble sized Gobabeb Gravel Formation in the valley upstream of Gobabeb prior to 9600 BP (Ward 1984).

2.4 CLIMATE OF THE NAMIB

The climate of the Namib is arid to hyper-arid but, especially in coastal areas, relatively cool. To the north, it grades into the summer rainfall desert of Angola; whilst to the south, rainfall occurs mostly in winter. A major feature of the region is the steep climatic gradient from the cool, foggy hyper-arid coastal zone to the hotter inland areas towards the Great Escarpment which receive scant summer rainfall (Besler 1972; Seely 1978).

The aridity of the climate results primarily from the latitudinal position of the region and the dominant effects of subtropical anticyclonic cells, especially that situated over the South Atlantic Ocean at 30° in summer, on the regional circulation pattern (Schulze 1972). In the central Namib, moist air masses can penetrate the area only when this anticyclonic cell is weak. However, their effects are limited, as the moist air is derived from the Indian Ocean and has to cross the subcontinent to reach the Namib. Thus descending, divergent air masses tend to occur all year. The effects of the subsistence induced stability are reinforced by the presence of the cold Benguela Current offshore, which intensifies the temperature inversion.

Rainfall in the central Namib and hence most of the northern and central parts of the sand sea falls during the summer months of January to April. The southern parts of the sand sea are marginal to the winter rainfall zone. Rainfall amounts are highly variable and localised spatially (Sharon 1981). Mean annual rainfall increases from 15 mm or less at the coast to 27 mm at Gobabeb and 87 mm at Ganab, near the escarpment (J.Lancaster et al. 1984). In the southern part of the sand sea, mean annual rainfall increases from 19 mm at Luderitz on the coast, to 89 mm at Aus in the escarpment zone (Royal Navy and South African Air Force 1944). From north to south, the annual total and proportion of winter rainfall increases. Along the coast, mean annual rainfall rises from 15 mm at Walvis Bay to 19 mm at Luderitz with a maximum in the period February to May. Alexander Bay and Port Nolloth receive 53-60 mm per year with 50-60% falling in the period April to August.

Advective fogs are a distinctive feature of the Namib climate. They occur as a result of the cooling of moist oceanic air as it passes over the Benguela Current. Their effects are felt for over 100 km inland. Days on which precipitating fogs occur decrease from 65 per year at the coast, to 37 at Gobabeb and 15 or less along the eastern edge of the desert. Fogs are most common in the winter at the coast, but in September to December inland. Fog precipitation is a significant moisture source for the Namib biota (Seely 1978) and may contribute to weathering and mineral breakdown (Goudie 1972; Sweeting and Lancaster 1982). The amount of fog precipitation rises from the coast where it averages 34 mm per year to a maximum of 184 mm 35-60 km inland and decreases sharply thereafter to 31 mm at Gobabeb and 15 mm to the east of the sand sea (J.Lancaster et al. 1984).

Temperatures in the central Namib are moderate by comparison with many other desert regions, reflecting the influence of the cold ocean offshore. Mean annual daily maximum temperatures range from 17°C at the coast to 28-33°C inland. Maximum daily temperatures of 38-40°C occur inland in February to March. Minimum daily temperatures average 13-16°C throughout the region and range seasonally from 4-8°C in June to August to 15-18°C in January to March.

Relative humidity in the Namib is strongly influenced by distance from the coast. Mean annual relative humidity falls from 87% at Walvis Bay to 50% near Gobabeb and 37% in the east. Periods of very low relative humidity (< 10%) are rare and occur when winter easterly 'berg' winds blow. Evaporation rates are high. Mean annual pan evaporation at Gobabeb is 3158 mm with a peak in December-January.

The circulation patterns of the central and southern Namib are strongly influenced by the South Atlantic anticyclone, situated offshore at 30°S (Schulze 1972). Overlain on the pattern of stable outblowing winds from the South Atlantic high pressure cell are the effects of local topographically and thermally induced circulations (Tyson and Seely 1980). Thus, throughout the year, southerly winds outblowing around the South Atlantic anticyclone are diverted inland as a SSW-SW sea breeze by the thermal contrast between the desert and the cold upwelling ocean waters. These winds reach their maximum frequency and strength in early to mid summer (September-January), when the anticyclone is at its strongest and most persistent and the thermal contrast between land and sea greatest (J.Lancaster et al. 1984). The overall strength of the southerly to south westerly winds decreases from south to north and from the coast inland. In winter (April to August), the South Atlantic anticyclone weakens and moves south-westwards, so that regional pressure gradients decrease and the continental, or South African anticyclonic cell, which is at its strongest at this time of year, exerts a major influence on circulation patterns. The southerly to south westerly sea breeze circulation is at its weakest and its inland penetration slight during winter months. Topographically induced mountain-plain

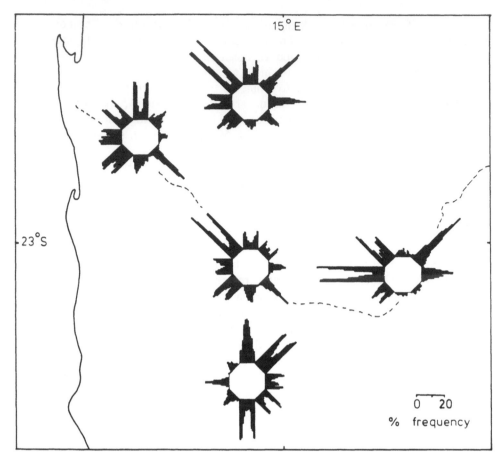

Figure 5. Seasonal variation of wind directions in the central Namib Desert. Wind stars show proportion of winds from major compass directions in each month, clockwise from January to December (after Lancaster et al. 1985).

and down valley winds are common during winter mornings, especially on the eastern margins of the desert (Tyson and Seely 1980). Of considerable importance in sand movements are the infrequent, but high velocity, easterly 'berg' winds which occur when pressure gradients are normal to the coast. At this time of the year southern areas of the Namib also experience north westerly winds associated with the passage of mid latitude cyclones to the south of the subcontinent. A further element in the regional circulation is introduced by the passage of shallow low pressure cells along the coast, especially in summer. As these pass southwards, a period of increased strength of southerly winds is followed by a few days of northerly winds.

Winds on the coast tend to be from SSE to SSW throughout the year (40-50% of all winds), with northerly winds in summer (8-10% of all winds) and, in winter, easterly 'berg' winds which reach the coast on only 10-15 days a year. Inland, especially in the central Namib (Fig. 5), a three component wind regime occurs: with a SSW-SW sea breeze which reaches a maximum frequency in summer (30-40% of winds in December) and is at a minimum in winter; a N-NNW plain-mountain wind on summer mornings (7-30% of winds) and an E-ENE mountain-plain wind which occurs on winter mornings and is responsible for up to 60% of winds at this season along the eastern edge of the desert. In addition, there are 'berg' winds in the winter, the frequency and persistence of which decreases from east to west.

At all seasons, there is a regular daily cycle of wind direction and velocity (Tyson and Seely 1980) as indicated by Figure 6. In the summer, mornings are calm or with a light northerly wind. These conditions last to 9-10h00 on the coast and midday or early afternoon inland. Thereafter, winds strengthen and back to SW or SSW, with a peak velocity being reached between 15h00 and 17h00. Evenings are a period of declining wind velocities. In winter months, the situation is similar on the coast, but the sea breeze is weaker. Inland on winter mornings winds are light to moderate easterly to north easterly, with topographic funnelling of winds down valleys leading from the escarp-ment. Later in the day, a period of calms or light northerly winds prevails until the south westerly sea breeze penetrates in the mid afternoon, often only for an hour or two. When 'berg' winds occur, they begin as SE or ESE winds in the early hours of the morning and back to NE or ENE as they rise in velocity to a peak during the late morning. On such days, afternoons are often calm, with a light SW breeze after 15h00.

There are important regional and seasonal variations in wind energy in the Namib. Mean annual wind velocities are highest at 6 m.sec⁻¹ or more in southern parts of the region and along the coast. They decrease inland to 3-4 m.sec⁻¹ on the eastern edge of the sand sea. Of greater importance is the proportion of the time the wind is blowing above

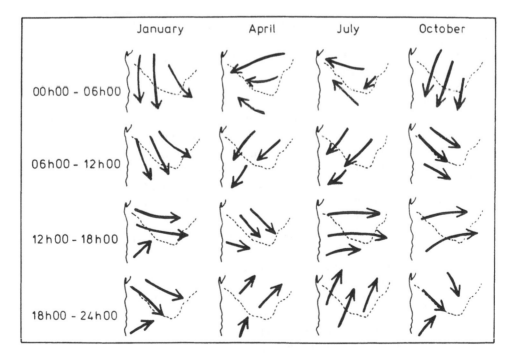

Figure 6. Daily cycle of wind direction in the central Namib Desert.

the threshold velocity for sand movement (4.5 m.sec⁻¹). In southern coastal areas, winds are above the threshold velocity for over half the time, reaching a peak of 70-80% of the time in December or January. Inland, the percentage of the time winds are able to move sand declines sharply to 25-35% in most areas and to around 20% along the eastern margins of the desert.

There is a clear seasonal pattern in the availability of wind energy to move sand. It is greatest in the period September to January, when winds are able to move sand for more than 30-40% of the time, and least in April-May, when winds are able to move sand for only 15-20% of the time. There is a secondary peak in wind energy in June or July, coinciding with the period when 'berg' winds occur.

2.5 ORIGIN OF THE PRESENT CLIMATIC PATTERN

A major cause of the current hyper-arid to arid climate of the central part of the Namib Desert is the northward flowing Benguela current system. The upwelling of cold water which originates in the sub Antarctic zone of the South Atlantic Ocean reinforces the subsistence induced stability of the anticyclonic circulation and influences the formation of advective fogs which contribute to the strong west-east climatic gradient. Thermal contrasts between sea and land generate local onshore sea breeze circulation patterns (Tyson and Seely 1980).

The origin of the Benguela Current and the onset of upwelling are often held to be synonymous with the origins of aridity in the Namib. Recently Ward et al. (1983) have suggested that desertic conditions probably existed in the central and northern Namib prior to the late Miocene. They argue that the present hyper-aridity is a late Cenozoic phenomenon and directly linked to the origins of upwelling. Although a northward flowing cold current may have existed from the Oligocene onwards (Van Zinderen Bakker 1975), the origins of upwelling are related to tectonic and oceanographic changes in the Southern Oceans leading to the development of the Antarctic ice caps (Kennett 1980; Van Zinderen Bakker 1975, 1984a) and the cooling of the Southern Ocean. These began in the late Oligocene (22-25 Ma) with the opening of the Drake Passage between Antarctica and South America, which allowed uninterrupted flow of the Circum Antarctic Current, and resulted in the thermal isolation and subsequent large scale glaciation of the Antarctic continent during the middle Miocene (12-14 Ma) (Kennett 1980). Cold sub-Antarctic water was carried northwards along the west coast of southern Africa by the South Atlantic gyre and resulted in the cooling and aridification of the subcontinent (Tankard and Rogers 1978).

Seisser (1978, 1980) has put forward a variety of sedimentary, palaeontological and geochemical data from DSDP Core 362 off the Walvis Ridge which suggests that, from the Oligocene to the mid Miocene, upwelling was weak and spasmodic. On land the Arriesdrift and Luderitz faunas suggest relatively mesic conditions in the mid Miocene with the southern Namib being covered by savannas with gallery forests along water courses (Corvinus and Hendey 1978). In the early late Miocene (c 7-10 Ma), conditions changed markedly. Warm water calcareous nannofossils and foraminifera were replaced by cold water species. Diatom production and sedimentation rates increased and the organic carbon and phosphorous content of the sediments rose sharply. Seisser attributes these changes to an intensification of upwelling from this time, and suggests that aridification of the Namib followed progressively in the Pliocene and Quaternary (Seisser 1978, 1980; Tankard and Rogers 1978). This was probably aided by global cooling from the late Miocene to early Pliocene (4-6 Ma).

Continued late Cenozoic aridity in the region is supported by data from pollen in deep sea cores (Van Zinderen Bakker 1984a) which indicates hyper-arid conditions throughout the Pliocene and Pleistocene and by the surface survival of calcrete palaeosols (Yaalon and Ward 1982) and lacustrine carbonates (Selby et al. 1979). Ward et al. (1983) and Lancaster (1984b) have suggested that Quaternary climatic fluctuations in the Namib were of low amplitude, compared to elsewhere in southern Africa and have been superimposed on a hyper-arid to arid mean. There is no evidence to suggest that the Namib has experienced any climate wetter than semi-arid at any time during the Quaternary.

It appears that the fluctuations in climate which have occurred have been superimposed on a trend to increasing aridity. Fluvial deposits from the Uis, Kuiseb and Tsondab rivers (Korn and Martin 1957; Ward 1984; Lancaster 1984a) all record a decline in average clast size from cobbles, gravels and pebbles to sands and silts during the Pleistocene, suggesting a parallel decline in discharge and stream competence.

3 DUNE MORPHOLOGY AND MORPHOMETRY

The pattern of dunes of different types and the spatial variation in their size, spacing and alignment in a sand sea may be regarded as the surface expression of the factors which control its dynamics and accumulation.

Three major dune types, classified according to the scheme of McKee (1979), occur in the Namib Sand Sea: crescentic, linear and star, in simple, compound and complex varieties (Fig. 7 and Photo 1). There are also small areas of zibar, sand sheets and shrub coppice dunes. The relative areal extent of each dune type in the sand sea is shown in Table 1. The following descriptions of dune morphology are based on studies of aerial photographs and satellite images, together with ground observations throughout most of the sand sea. Measurements of dune width and spacing were made from aerial photographs, whilst measurements of dune height were obtained during field survey at the sediment sampling sites (Fig. 21).

3.1 CRESCENTIC DUNES

Crescentic and barchanoid dunes of simple and compound form (types 1a-c on Fig. 7) cover 13.3% of the area of the sand sea. They occur in a strip up to 20 km wide along the coast and are also present further inland where dune patterns are disturbed by river valleys, for example at Sossus Vlei and on the Tsondab Flats.

There are two major areas of crescentic and barchanoid dunes. The southern group extends from Elizabeth and Chamais Bays to Sylvia Hill. It begins as a narrow train of irregularly spaced barchans and barchanoid ridges which extends north from the wide Atlantic beaches towards Kolmanskop. These dunes constitute the major present day corridor for sand input to the sand sea (Rogers 1977). North of Luderitz, an extensive area of simple and locally compound crescentic and barchanoid ridges fans out into the main sand sea. Inland, these dunes merge with low, compound linear dunes on SE-NW alignments. To the west, the simple crescentic ridges feed a belt up to 10 km wide of compound crescentic dunes which extends up the coast as far as Sylvia Hill.

The northern group of crescentic dunes begins on the coast south east of Meob Bay and continues up the coast to the northern boundary of the sand sea at the Kuiseb delta as a 20 km wide belt of large compound crescentic dunes. Smaller areas of low crescentic and barchanoid ridges, with isolated barchans in places, occur on the coastal flats and deflated

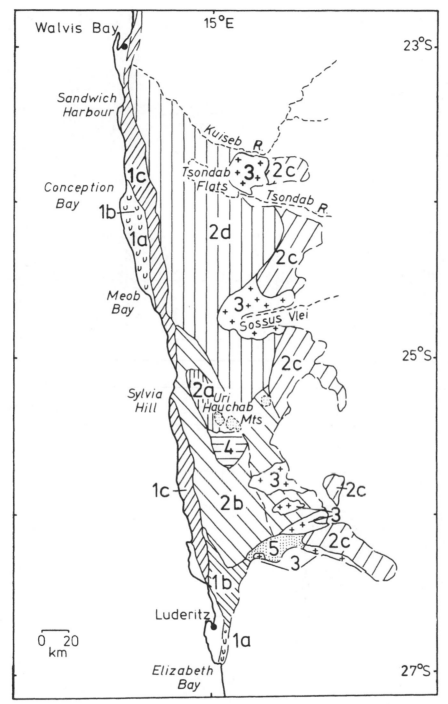

Figure 7. Distribution of different dune types in the Namib Sand Sea. Areas of shrub coppice dunes along coast not shown. 1: Crescentic dunes; a) Barchans; b) Simple crescentic dunes; c) Compound crescentic dunes; 2: Linear dunes; a) Simple; b) Compound, straight; c) Compound, anastomosing; d) Complex; 3: Star dunes and chains of star dunes; 4: Large zibar; 5: Sand sheets.

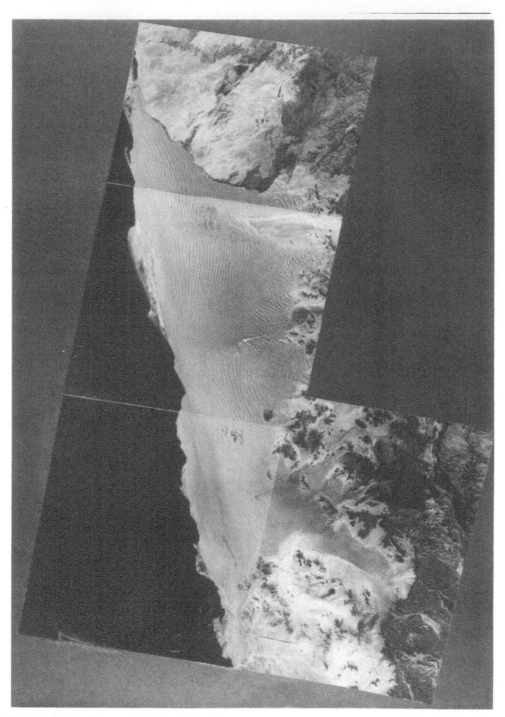

Photo 1. Landsat Image of the Namib Sand Sea.

22

Table 1. Relative areal importance of dune types in the Namib Sand Sea.

Dune type	Area covered (% of sand sea)
Linear	74.0
Simple	2.1
Compound	
Straight	19.6
Anastomosing and reticulate	15.5
Complex	36.9
Crescentic and barchanoid	13.3
Barchans	0.1
Simple	3.5
Compound	9.8
Star dunes and chains of star dunes	9.3
Large zibar	1.5
Sand sheets	3.3
Shrub coppice dunes	0.1

bedrock terrain between Meob and Conception Bays. Some of these dunes can be seen to join the main dune mass from the west.

3.1.1 *Morphology of crescentic dunes*

Barchans are rare in the Namib Sand Sea. Individual dunes and trains of barchans cover less than 0.1% of its area. They occur mainly in peripheral areas of the sand sea, particularly in upwind, near source, zones and where topography is irregular. Examples of such areas are the sand stream which feeds the sand sea south of Luderitz and the Conception-Meob area (Photo 2). Barchans also occur on the downwind margins of the sand sea, beyond the advancing tips of linear dunes on the Tsondab Flats (Lancaster 1980) and east of Gobabeb, as well as in the Kuiseb delta (Barnard 1975). Large compound barchans, similar to the Pur Pur dune of Peru (Simons 1956) were observed in some areas south of Luderitz. Most barchans are less than 5 m high and irregularly spaced.

Simple crescentic ridges with straight or slightly sinuous crestlines and simple barchanoid ridges, with connected crescentic slip faces cover some 3.5% of the area of the sand sea. They are found mostly in the area northeast of Luderitz and in coastal areas near Conception Bay (Photo 2) and Sandwich Harbour. There are also small groups of crescentic and barchanoid ridges in the area of disturbed dunes west of the Tsondab Flats.

In the Namib Sand Sea, simple crescentic dunes occur mainly in the transition zones between irregularly spaced barchans and the large compound crescentic ridges. In these areas, sand cover is often semi complete, with small windows of the sub dune surface exposed. In many places, there are extensive areas of coarse mega rippled sands in interdunes between simple crescentic ridges.

Most simple crescentic and barchanoid ridges are between 3 and 10 m high, with a

Photo 2. Barchans and simple crescentic dunes southeast of Conception Bay. Note large expanses of sub dune bedrock surface.

Photo 3. Compound crescentic dunes south of Sylvia Hill. Note two orders of bedform spacing and superimposed dunes in crestal areas.

24

spacing of 100-400 m. In profile, windward slopes steepen from 2-3° at the base to 10-12° in the upper mid slope and decline again to 2-3° at the crest, which is generally sharp, but occasionally rounded, with the brink of the slip face, at an angle of 32-34° lying beyond the crestline.

Simple crescentic ridges in the Namib Sand Sea are comparable in size with those at White Sands, New Mexico (McKee 1966; Breed 1977), the Guerro Negro, Baja California (Inman et al. 1966) and the Skeleton Coast dunefield in the northern Namib (Lancaster 1982a).

Compound crescentic dunes occupy some 10% of the area of the Namib Sand Sea. They usually consist of a main dune ridge 10-40 m high with a spacing of 800-1200 m. On their upper stoss slopes and crests there are often small barchanoid ridges 2-5 m high and 50-100 m apart (Photo 3). When they extend to the lee side of the crest the barchanoid ridges form a series of 3-5 m high descending slip faces. In some areas, the crestline of the main dune consists in plan of a series of deeply concave crescentic slip faces 10-20 m high, backed by small barchanoid dunes, which are separated by areas where the crestline is gently rounded. In inland areas of the northern belt of compound crescentic dunes, there is often a prominent linear element on a north south alignment, which crosses the interdune between adjacent main dune ridges (Photos 3 and 4). Many of these linear ridges can be considered to be incipient linear dunes.

Compound crescentic dunes occur where sand cover is complete or near complete. There are no true interdune areas and only rarely is the sub dune surface exposed in deep hollows at the base of strongly concave slip faces. Generally the lee face of one dune abuts the lower parts of the stoss slopes of the next dune downwind. Throughout the area

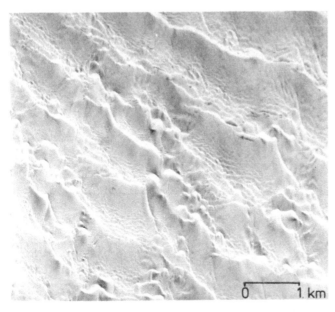

Photo 4. Vertical aerial photograph of compound crescentic dunes with oblique linear ridges (?incipient linear dunes). East of Conception Bay.

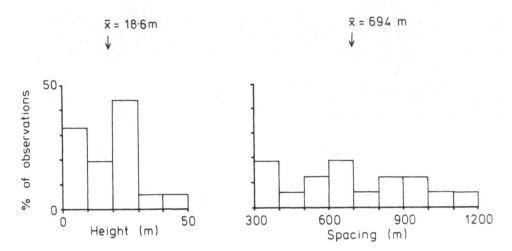

Figure 8. Morphometry of crescentic dunes.

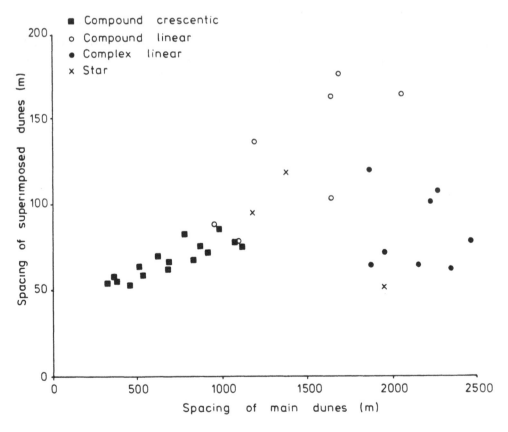

Figure 9. Relationships between the spacing of main and superimposed dunes of each major type.

of crescentic dunes of all varieties there is very little vegetation. Occasional plants of *Trianthema hereroensis* and *Stipagrostis sabilicola* may be found, at the base of slip faces and on the lower parts of stoss slopes.

Compound crescentic dunes in the Namib Sand Sea are similar in size to those identified as simple varieties by Breed and Grow (1979) in Al Jiwa, the United Arab Emirates and Takla Makan and to those identified as compound in the Tunisian, KaraKum and Gran Desierto sand seas. They are also very similar in size to the dunes of the Algodones dunefield, California (Norris and Norris 1961), but lack the multiple superimposed bedforms developed on these dunes.

3.1.2 *Morphometry of crescentic dunes*

The morphometry of crescentic dunes in the Namib Sand Sea is summarised in Table 2 and Figure 8. Their spacing is variable and ranges from 100-1400 m, with a mean of 610 m. Most crescentic dunes are spaced 200-1000 m apart. Throughout their area of occurrence, the spacing of crescentic dunes is much less regular than that of other dune types and the pattern seems much less well organised.

Crescentic dune ridges with a spacing of 400 m or less are usually simple varieties, with a height of 5-10 m. Larger and more widely spaced crescentic dunes are invariably of compound form, with the main ridges spaced at 600-1000 m and the crestal elements at 50-100 m, or about one tenth that of the main dunes (Fig. 9). There is a strong correlation between the spacing of the main ridge and that of the superimposed bedforms of crescentic dunes (r = 0.84, significant at the 0.05 level), but not for other dune types, as Figure 9 shows. Spacings of the southern group of compound crescentic dunes range between 600 and 1000 m with a mean of 792 m. Two modes of spacing, at 600-800 m and 900-1100 m, can be distinguished for the northern group of compound crescentic dunes. Heights of compound crescentic dunes range between 15 and 40 m.

There is a close relationship (Fig. 10) between the height and spacing of all varieties of

Table 2. Crescentic dune morphometry at sample sites (mean values).

Site	Height (m)	Spacing (m)
IXa	10	295
IXb	15	500
IXc	10	300
XI	27	442
XVI	26	1063
XXIa	46	1050
XXIb	26	1097
XXIc	18	858
XXId	7	220
XXIIa	5	150
XXIIb	18	760
XXIIIa	5	275
XXIIIb	10	250
XXIIIc	4	190

crescentic dunes (r = 0.75, significant at the 0.05 level). Similarly close relationships occur for crescentic dunes in the Skeleton Coast dunefield and the Gran Desierto (Lancaster 1982a; Lancaster et al. 1987).

3.2 LINEAR DUNES

Linear dunes (type 2a-c on Fig. 7) on S-N to SE-NW alignments are the dominant dune form in the Namib Sand Sea and cover 74% of its area. Three varieties can be recognised. Simple linear dunes, with a single narrow sharp crested ridge are rare and occur over only 2% of the sand sea, mostly in its southern and eastern parts. Compound linear dunes cover 35% of the sand sea area in its southern and eastern sectors. Two sub varieties occur: straight, parallel dunes (type 2b) with 2-5 parallel or converging crestal ridges on a broad swell are common in the southern part of the sand sea (20% of its area); and those with an anastomosing or reticulate pattern of crestal ridges (type 2c), partly to largely fixed by vegetation, which occur along the eastern margin of the sand sea (15% of its area). Large complex linear dunes (type 2d), with stellate peaks at intervals and barchanoid dunes on their eastern flanks dominate in central and northern parts of the sand sea and are found over 37% of its area. They are transitional to chains of star dunes in some areas.

3.2.1 *Morphology of linear dunes*

Typical compound linear dunes in the southern parts of the Namib Sand Sea consist of 3 and locally up to 5 slightly sinuous 5-10 m high sharp crested ridges running parallel or

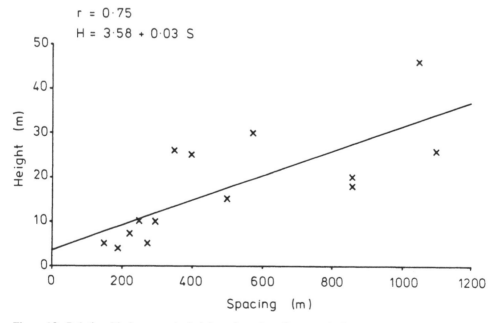

Figure 10. Relationship between the height and spacing of crescentic dunes.

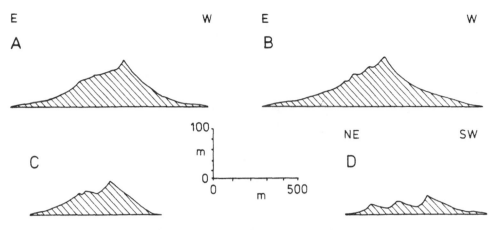

Figure 11. Surveyed profiles of complex (A-C) and compound (D) linear dunes.

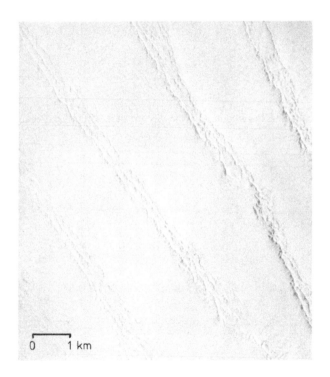

Photo 5. Compound linear dunes in the southern part of the sand sea.

sub parallel to each other on SE-NW alignments (Photos 4 and 5). They lie on a 500-800 m wide gently convex linear ridge (Fig. 11D), composed of coarser sand and with the same general alignment as the crestal ridges. This is comparable with the 'whalebacks' of Bagnold (1941) and the 'ondulations longitudinales' of Mainguet and Callot (1978). Total height of these dunes is 25-50 m and they have a spacing of 200-2000 m. Locally

29

Photo 6. Partly vegetated reticulate compound linear dunes in the eastern part of the sand sea.

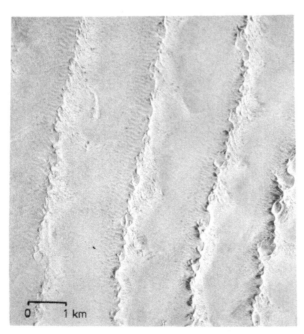

Photo 7. Aerial view of complex linear dunes in the central part of the sand sea. Each ridge is approximately 150 m high. Note star dune peaks at intervals and small east flank barchanoid dunes.

30

the crestal ridges may join in Y junctions open to the south or south east. In other places, dendritic or 'feather barb' patterns may occur.

The crestal ridges may be continuous, especially on the south western side of the dunes, but discontinuous in the centre and on the eastern side of the dunes. The ridges are asymmetric, with the aspect of the slip faces, at angles of 32-34°, changing seasonally from SW or W in April to September and NE or E during the remainder of the year. Windward slopes of the crestal ridges are between 15 and 25°. The lightly vegetated lower slopes, or plinths, on each side of the dune slope at 2-6° steepening towards the centre of the dune. Interdune areas between compound linear dunes of this type are generally sand covered and formed into low undulations, comparable with the 'zibar' of Holm (1960), which continue onto the adjacent plinths.

Compound linear dunes in the southern parts of the Namib Sand Sea are directly comparable with the 'bouquets de silks' and 'silks sur ondulations' described by Mainguet and Callot (1978) from the Erg Fachi Bilma.

In the eastern parts of the sand sea, compound linear dunes on S-N or WSW-ENE alignments are common. They have an anastomosing or reticulate pattern of crestal ridges and prominent 'feather barb' junctions (Photo 6). They correspond to the dunes described as 'complex multicyclic' by Barnard (1973) and 'lace dunes' by Besler (1980). Many of these dunes are partly stabilised by a discontinuous cover of grasses, mainly *Eragrostis spinosa* and *Stipagrostis sabilicola* such that only crestal areas are active.

Most complex linear dunes in the Namib Sand Sea consist of a single main dune ridge on a S-N to SSE-NNW alignment which rises to 50-170 m above adjacent interdune areas (Photos 7 and 8).

The crestline of the main ridge is sharp and sinuous and connects a series of regularly spaced peaks which, especially in the central parts of the sand sea, have a stellate form. This gives the ridge a serrate appearance. The peaks often occur at the westward turn of the crestal sinuosity and effectively form a series of reversing dunes along the main crestline. Locally there are intersecting ridges and deep blow out hollows adjacent to the dune peaks.

The major slip face, at an angle of 32-33°, faces east or north east and may be 10 m high at the time of its maximum development in March. In winter, east to north east winds erode its upper section and reverse the orientation of the slip face to face west or south west. However, this slip face is rarely more than 5 m high.

Below the slip faces on the eastern side of the dune is a wide, gently sloping plinth, lightly vegetated with clumps of *Stipagrostis sabilicola* and *Trianthema hereroensis* (Robinson and Seely 1980). *S.sabilicola* is more common in eastern parts of the sand sea, whilst *T.hereroensis* is dominant to the west. Secondary or superimposed dunes of barchanoid or crescentic form, 2-10 m high and 50-200 m apart, are developed on the upper parts of the plinth (Photo 9). In many instances, they join the main crestal ridge at the westerly turn of its sinuosity and so create an en echelon pattern of crests. Strong development of east flank barchanoid dunes seems to be associated with a sinuous main crest. Where it is straight, or slightly sinuous, such dunes are absent or poorly developed. Particularly in western parts of the central and northern sand sea, the eastern ends of the east flank dunes may be recurved at irregular intervals to join with corridor crossing dunes of simple linear form (Photo 10).

Western slopes and plinths of the linear dunes are smooth or gently undulating in a

Photo 8. Ground view of western slope of large complex linear dunes. Note regular spacing of dune peaks.

Photo 9. Barchanoid dunes on east flank of a complex linear dune.

Photo 10. Simple linear dune on WSW-ENE alignment crossing corridor between complex linear dunes. View to west.

direction normal to the main dune trend. Slope angles steepen from 2-5° on the plinth to 15-20° near the crest. As shown by the profiles in Fig. 11, the eastern slopes of the dunes, apart from the slip face, are less steep, with angles of 2-5°.

Apart from some areas in the northern parts of the sand sea adjacent to the Tsondab and Kuiseb valleys interdune areas between complex linear dunes are sand covered and covered by zibar on a trend normal or slightly oblique to that of the main dune ridges. In places, these undulations continue onto the plinths and upper western slopes of the adjacent linear dunes.

Variants of the complex linear dunes are of two main types. In eastern parts of the sand sea, north of Tsondab and Sossus vleis, complex linear dunes merge with chains of star dunes (compound star dunes). Elsewhere, towards their eastern limits, the complex linear dunes become lower and broader. The main ridge becomes less prominent and barchanoid and reversing elements more pronounced, eventually resulting in a reticulate pattern of crestal ridges and a compound linear dune. Another instance of a transition between complex and compound dunes occurs in the area north of the Uri Hauchab Mountains, where the change from compound to complex forms is associated with the growth of the western ridge of a compound dune so that it dominates the dune and reduces the other ridges to small, often barchanoid forms. The complex linear dunes of the Namib Sand Sea are similar to the ''uruq' of the Rub al Khali described by Holm (1960) and to some of the 'chaines ghourdiques' and 'ghourdes en chaines' of Mainguet and Callot (1978).

33

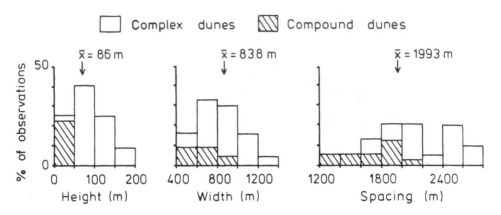

Figure 12. Morphometry of compound and complex linear dunes.

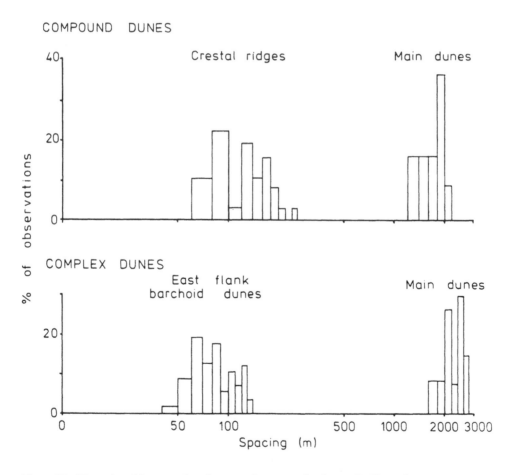

Figure 13. Hierarchy of dune spacings in areas of compound and complex linear dunes.

Table 3. Morphometry of linear dunes at sample sites (mean values).

Site	Height (m)	Width (m)	Spacing (m)
I	65	825	1900
II	101	1037	2565
III	78	688	1050
V	76	812	1500
VI	127	919	1837
VII	139	818	2520
VIII	161	904	2535
IX	44	600	1500
XII	56	665	2130
XIII	74	857	2055
XV	40	656	2004
XVII	27	400	1185
XVIII	72	1125	1900
XXV	35	707	1733
XXVI	35	878	1830

Other similarities are with dunes in the Wahiba Sands (Breed and Grow 1979) and with those called 'bras' by Capot Rey in the Erg Occidental and Mourzuk sand sea (Capot Rey 1945, 1947).

3.2.2 Morphometry of linear dunes

The morphometry of linear dunes at sample sites is summarised in Table 3 and Figure 12. Spacing of the linear dunes varies between 1200 and 2800 m. Mean spacing of compound dunes is 1724 m and most are 1600-2000 m apart. Complex dunes are more widely spaced, with a mean of 2108 m and a range of 1600-2800 m. A hierarchy of dune spacings exists in areas of both compound and complex dunes (Fig. 13). In areas of compound linear dunes, the spacing of the crestal ridges is generally between 120 and 180 m, or approximately one tenth that of the main dunes and varies with the spacing of the main dune. In areas of complex linear dunes, the two orders of spacings correspond to the main dune ridges and the east flank barchanoid dunes, which are spaced 60-140 m apart, with a mean of 87 m, but there is no relationship between the spacing of the main dune and the superimposed ridges (see Fig. 9).

The width of linear dunes ranges between 400 and 1400 m. Compound dunes are generally narrower than complex forms, with a mean width of 665 m, compared with 886 m for complex linear dunes. Linear dune height ranges from 25 to 170 m. Compound varieties are relatively low, with a mean height of 35 m, compared to a mean of 99 m for complex dunes. Most compound linear dunes are 25-45 m high and complex varieties are generally 80-120 m in height.

The height, width and spacing of linear dunes are closely interrelated, as Figure 14 shows. Linear dune height and spacing are directly correlated ($r = 0.75$, significant at the 0.05 level). As observed by Breed and Grow (1979), dune width and spacing are also correlated ($r = 0.46$).

3.3 STAR DUNES

Star dunes (type 3 on Fig. 7) are found in groups along the eastern margins of the sand sea. The largest cluster is centered on Sossus Vlei. Other important areas of star dunes occur north of Tsondab Vlei and along the eastern margins of the sand sea south of Nam Vlei, where narrow 'fingers' of star dunes extend for up to 20 km west into the sand sea (Photo 1). Star dunes are also present locally along the Koichab valley on the southern

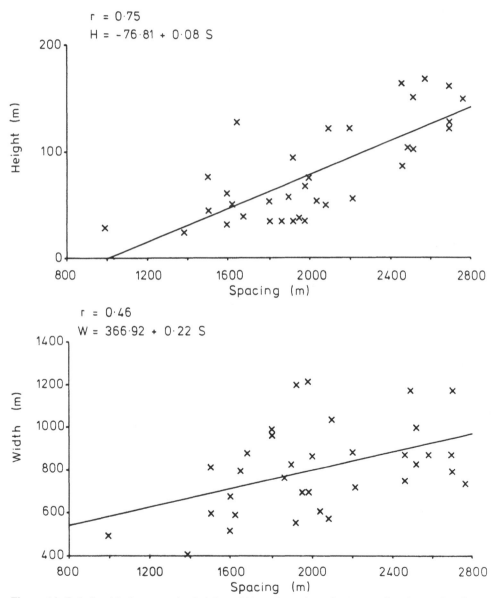

Figure 14. Relationship between the height, width and spacing of compound and complex linear dunes.

36

margins of the sand sea and adjacent to the Kuiseb River. In total, star dunes cover some 9% of the sand sea area.

3.3.1 *Morphology of star dunes*

Few of the dunes identified as star types in the Namib Sand Sea have a truly stellate form, with arms radiating from a pyramidal shaped central peak. Many consist of a single narrow, relatively short (0.5-2 km), steep sided ridge with a straight or sinuous crestline (Photo 11). In profile (Fig. 15) the ridge is near symmetrical. Many ridges have a strongly prefered alignment, usually SE-NW or SSE-NNW. From the crestal ridge, curving arms descend on alignments which are roughly perpendicular to the crest. Dunes of this type

Photo 11. Chains of star dunes north of Tsondab Vlei. View to north.

Photo 12. Pyramidal isolated star dune in southeastern part of the sand sea.

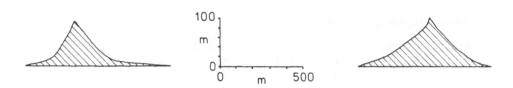

Figure 15. Surveyed profiles of star dunes.

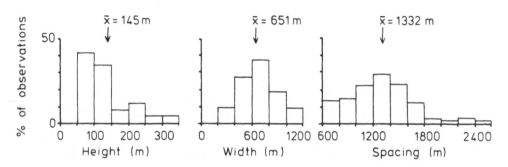

Figure 16. Morphometry of star dunes.

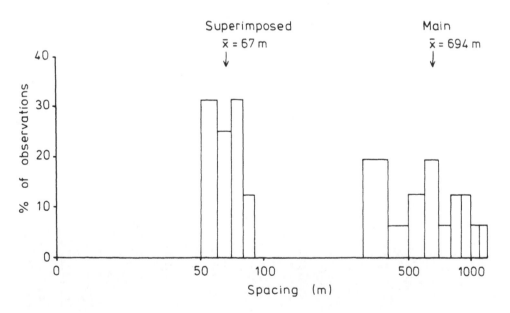

Figure 17. Hierarchy of dune spacings in areas of star dunes.

are the largest in any area and in the Namib Sand Sea as a whole, reaching heights of 200-350 m around Sossus Vlei. In this area, many star dunes coalesce to form chains of star dunes, similar to the 'chaines ghourdiques' of Mainguet and Callot (1978) and the chains of star dunes in the Gran Desierto, Mexico described by Lancaster et al. (1987).

Star dunes in the southern parts of the sand sea tend to have a more pronounced stellate or pyramidal form (Photo 12) and are similar to isolated dunes in the Gran Desierto, Mexico and in Niger (Breed and Grow 1979). They tend to be much lower, at 80-100 m high, than the elongate ridge variety.

Slip faces develop on both sides of the crest and the subsidiary arms, their orientation depending on the winds of the time. They face NE to E in summer and SW to W in winter. In some southern parts of the sand sea, slip faces with a SE or SSE orientation may also develop in winter.

The lower slopes of the star dunes form a wide, undulating plinth with slope angles of 2-5°. In many areas, small barchanoid and reversing dunes develop and merge with the arms descending from crestal ridges. Plinths of star dunes are often quite well vegetated, with large clumps of *Stipagrostis sabilicola* and *Eragrostis spinosa*. There are often deep hollows and blowouts between and adjacent to the arms descending from the crestal areas. Interdunes between star dunes are often irregular and quite well vegetated, with areas of small barchanoid and reversing dunes at intervals.

3.3.2 *Morphometry of star dunes*

Figure 16 and Table 4 summarise the morphometry of star dunes in the Namib Sand Sea. The spacing of star dunes and related chains of star dunes varies very widely, from 600-2600 m, but most star dunes are between 1000 and 1800 m apart, with a mean of 1332 m. As with other large dune types a hierarchy of spacings occurs, with small barchanoid and reversing dunes with a spacing of 90-120 m on plinths and in interdune areas (Fig. 17).

Widths of star dunes lie between 400 and 1000 m, with a mean of 651 m. Star dunes of the type found in the Namib Sand Sea appear to be narrower and more closely spaced than complex linear dunes of a comparable size. The mean height of star dunes is 145 m, which is rather greater than that for complex linear dunes. The largest star dunes occur in the vicinity of Sossus Vlei, where they attain heights of 300 m or more. Those in the southern parts of the sand sea are much lower at 80-150 m and more closely spaced (± 1000 m). There is a close correlation between star dune height and spacing, (r = 0.72, significant at the 0.05 level) as shown by Figure 18. Although Breed and Grow (1979)

Table 4. Morphometry of star dunes at sample sites (mean values).

Site	Height (m)	Width (m)	Spacing (m)
X	186	765	1590
XIV	91	540	1110
XIX	92	621	1642
XXIV	106	780	1456
XXVa	80	360	720

could find no statistically significant relationship between star dune diameter and spacing, such a relationship does exist in the Namib Sand Sea (r = 0.68, significant at the 0.05 level) as demonstrated in Figure 18.

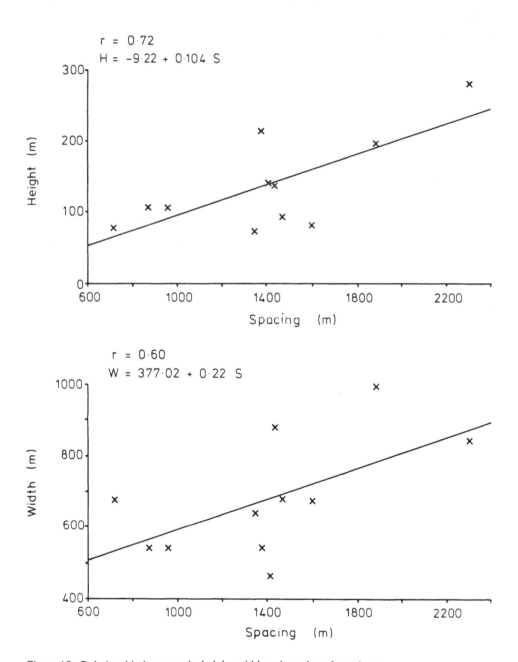

Figure 18. Relationship between the height, width and spacing of star dunes.

3.4 OTHER DUNE TYPES

In the Namib Sand Sea, areas without significant dune development are rare. On the southern margins of the sand sea, north east of Koichab pan, there is an area of gently undulating to flat sand sheets which grades northwards to the interdunes between compound linear dunes (Photo 13).

South and south east of the Uri Hauchab mountains there is an area of rolling sands

Photo 13. Partly vegetated sand sheet in southern part of sand sea. Star dunes on horizon to northeast.

Photo 14. Low rolling dunes without slipface development (giant zibar) south of the Uri Hauchab mountains (on horizon).

Photo 15. Shrub coppice dunes on coast south of Walvis Bay. View east towards compound crescentic dunes.

which are formed into 10-30 m high ridges with a broad rounded crest on E-W to ESE-WNW alignments (Photo 14). Despite the height of these dunes, slip faces are only developed in the northern parts of the area covered by these dunes. These low rolling dunes without slip faces are similar in size and spacing to some compound crescentic dunes and appear to be similar in form to dunes described and illustrated as 'mréyé' by Monod (1958). They also resemble very large zibar. To the north west, they form the basal areas of simple and compound linear dunes. On the southern margins of this dune type, the pattern of ridges appears to continue that of the zibar between the compound linear dunes which seem to be advancing northwards over the undulations.

Shrub coppice dunes (Photo 15), resulting from deposition around the roots of actively growing plants of *Salsola* sp. are found in coastal margin areas of the Namib Sand Sea. Their height ranges from 0.5 m to 3-5 m. As in the Skeleton Coast dunefield of the northern Namib, they are clustered along sand streams feeding the sand sea from beaches. Large areas of these dunes occur south of Walvis Bay and in the area east of Meob Bay.

3.5 THE PATTERN OF DUNE SIZE AND SPACING

Dune size (height, spacing) vary together in a systematic way in the Namib Sand Sea (Fig. 19). Dunes are highest and most widely spaced in the central and some northern parts of the sand sea, with progressively smaller and more closely spaced dunes towards the margins.

In the southern parts of the sand sea, most dunes are less than 50 m high, except for the

42

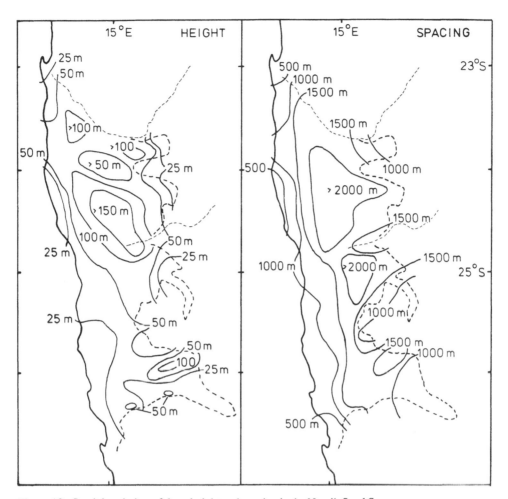

Figure 19. Spatial variation of dune height and spacing in the Namib Sand Sea.

groups of star dunes along the Koichab valley and the eastern margins. Dune height increases to a maximum of 180-200 m in the area between Sossus Vlei and the Tsondab valley. A smaller area of large linear and star dunes occurs between Tsondab Vlei and the Kuiseb River. Dune height decreases towards the eastern margins of the sand sea, where, except for areas of star dunes, most dunes are less than 25 m high.

Dune spacing varies in a similar way, with the most widely spaced dunes occurring in the central and northwestern parts of the sand sea and progressively more closely spaced dunes towards the margins. However, spacing of dunes in marginal areas, e.g. along the Kuiseb valley and the eastern edge of the sand sea, tends to be much greater than the height of the dunes would suggest. Within the area of crescentic dunes along the coast north of Meob Bay, there is a striking variation in dune size and spacing. Dune spacing increases northwards from 700 m east of Meob Bay to 1000-1200 m in the area northeast of Conception Bay and then declines progressively to 300-400 m at Sandwich Harbour.

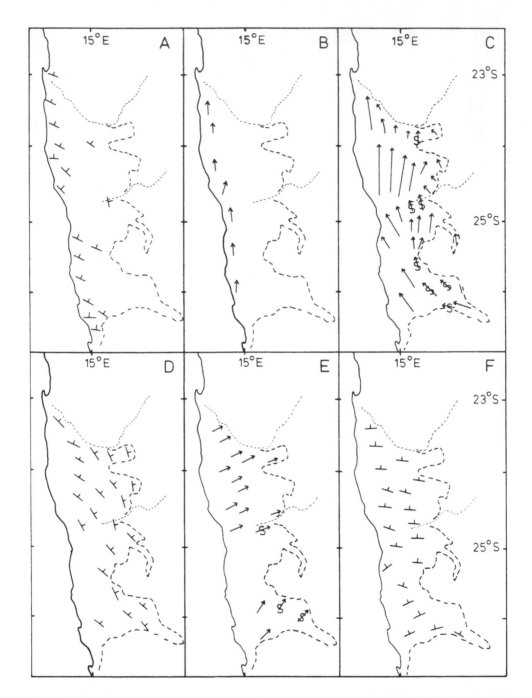

Figure 20. Pattern of dune alignments in the Namib Sand Sea. A: Crescentic dunes. Bar indicates strike of crest; tick, orientation of slip face; B: Oblique linear elements in areas of crescentic dunes; C: Linear dunes and main ridges of star dunes; D: East flank barchanoid dunes in areas of complex linear and star dunes. Bar indicates strike of crest; tick, slip face orientation; E: Corridor crossing dunes in areas of linear dunes and subsidiary arms of star dunes; F: Zibar, Bar indicates strike of crestline; tick, lee slope.

Dune height changes in a similar way, increasing from 10-18 m in the south to 40-50 m in central areas and falling northwards to 20-30 m at Sandwich Harbour and 5-15 m adjacent to the Kuiseb River.

3.6 THE PATTERN OF DUNE ALIGNMENTS

In many areas of the sand sea, the pattern of dune alignments consists of a major or dominant trend, together with one or more subsidiary elements. From place to place, the relative importance of these trends may vary and a major trend in one area dies out or is continued as a subsidiary element, whilst a minor trend is accentuated to dominate in another area. In this way, the trend of the crestal ridges of the coastal crescentic dune belt is continued inland by that of the barchanoid dunes on the east flanks of complex linear dunes (Fig. 20A and D).

The trend of the crests of most simple and compound crescentic dunes is 300-320°, with north east facing slip faces, and tends to swing round slightly inland. In the Conception-Meob area, simple crescentic ridges and barchans have ridge trends of 240-280° (Fig. 20A). Especially along the inland margin of their distribution, many compound crescentic dunes have prominent linear elements which cross from one dune to another on NNW-SSE to N-S alignments (Fig. 20B).

The major dune trend in the sand sea is that of the linear dunes (Fig. 20C), which are aligned NNW-SSE to NNE-SSW in central and northern areas. Compound linear dunes in southern parts of the sand sea are on NW-SE alignments. The trend of the anastomosing and reticulate compound dunes along the eastern margins of the sand sea is often irregular, as dunes are diverted around inselbergs, but prominent alignments on WNW-ESE to NW-SE are common.

Within areas of complex linear dunes, there are two subsidiary dune trends. Crests of east flank barchanoid dunes trend 320-330°, with NNE-NE oriented slip faces, in western areas, but swing round eastwards to 340-355°, with east facing slip faces (Fig. 20D). Corridor crossing dunes (Fig. 20E) follow a consistent WSW-ENE, or 060° trend in all areas where they occur. Low rolling dunes or zibar in interdune areas tend to have crest alignments normal to the main linear dunes (Fig. 20F).

The trend of the crests of star and reversing dunes follows that of the linear dunes (Fig. 20C). Subsidiary arms tend to follow the trend of corridor crossing dunes (Fig. 20E), whilst barchanoid dunes on their lower flanks continue the pattern of similar dunes elsewhere.

3.7 THE PATTERN OF DUNE MORPHOLOGY AND MORPHOMETRY IN THE SAND SEA

The morphology and morphometry of all dune types varies in a consistent way throughout the sand sea. Thus the largest and most widely spaced dunes are found in the central and northern parts of the sand sea, with progressively smaller dunes towards the margins and generally low dunes in its southern areas.

A line drawn from south of Meob Bay through the Uri Hauchab Mountains effectively divides the sand sea into two contrasting parts. The southern area is characterised by low compound linear dunes, with large areas of zibar and a belt of simple and compound crescentic dunes along the coast. Small tongues of star dunes penetrate the eastern margins of the sand sea. The northern and central section of the sand sea is dominated by large complex linear dunes, locally grading east and southeast into chains of star dunes and isolated star dunes. The northern margins of the sand sea east of Gobabeb are characterised by partly sand free interdune areas. Along the eastern margins is a wide zone of reticulate and anastomosing partly vegetated compound linear dunes. A belt of compound crescentic dunes extends up the coast from Meob Bay.

Similar 'facies sequences' of dune type, size and spacing have been reported from many other sand seas (Kadar 1934; Norris and Norris 1961; Wilson 1973; Breed et al. 1979; Porter 1986). The patterns are characterised by an increase in sand cover, dune size and complexity of dune patterns towards the centre of the sand sea. Following Porter (1986), sand seas may be divided into three zones: the upwind or 'back erg' area, with extensive areas of zibar and sand sheets left behind as the sand sea migrates downwind, together with scattered low dunes; the 'central erg', with large dune complexes representing the depositional centre of the sand sea; and the 'fore erg', characterised by low and/or widely spaced dunes, which migrate across the underlying non dune substrate. The pattern of dunes in the Namib Sand Sea appears to generally follow such a model.

4 DUNE SEDIMENTS

The Namib Sand Sea is composed dominantly of yellowish brown to yellowish red medium to fine (Wentworth scale) quartz sand. The sediments which are contained in the dunes of the sand sea, together with those of the interdune areas are the product of the ongoing process of sand sea accumulation. Their character provides information on how the sand sea has developed and the nature of the sedimentary processes involved.

Study of modern dune sediments in most desert sand seas is hampered by the absence of exposures of subsurface deposits and by the very considerable logistic problems of investigating sediments at depths greater than 1 m below the dune surface. Consequently, much of the material contained in this chapter concerns only the surface sediments of dunes and interdune areas, which were studied at 26 major sites throughout the sand sea (Fig. 21).

4.1 SAND COLOURS

Colours of sands in the Namib Sand Sea vary from light yellowish brown (10YR 7/4) through reddish yellow (7.5YR 5/6 to 5/8) to yellowish red (5YR 5/8). The distribution of colour follows three south to north zones (Fig. 22), with pale colours in the coastal zone, reddish yellow colours in the central parts of the sand sea and redder hues along the eastern margins. The pattern of colour change inland from the coast has been noted by many investigators in the region, most notably by Logan (1960), who attributed the change inland to the increasing age of the sands. Besler (1980) suggested that the observed change related to loss of red coatings as the sands were transported by fluvial processes from east to west.

Studies of individual sand grains under the microscope show that increased reddening of the grains is achieved by a greater extent and thickness of iron oxides deposited in pits and other surface irregularities on clear or frosted quartz grains. Reddening is most pronounced on smaller (2-3 phi), more angular grains, as was observed by Folk (1976) from Simpson Desert dunes. In some areas, especially toward the east of the sand sea, reddening is the result of the amber colour of many quartz grains.

The colour of Namib sands is relatively pale by comparison with dunes from the Kalahari and Simpson-Strzelecki deserts where sand colours are red to yellowish red (2.5YR 5/8 to 7.5YR 5/8) (Folk 1976; Wasson 1983b; Lancaster 1986). Sands from

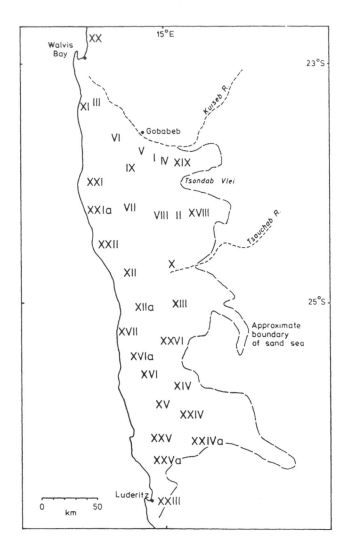

Figure 21. Location of sites in the Namib Sand Sea where dune sands were sampled.

eastern and central parts of the sand sea are comparable in colour with those in the north western Sahara (Alimen et al. 1957); Libya (Walker 1979) and Arabia (Besler 1982). Sands from coastal areas are similar in colour to other coastal dunefields in the Namib Desert such as that on the Skeleton Coast, where sands are very pale grey (10YR 7/3) to light yellowish brown in colour (10YR 6/4 to 7/4) (Lancaster 1982a).

4.1.1 *Spatial variation in sand colour*

It has been widely reported, for example by Alimen et al. (1958), Logan (1960), Wopfner and Twidale (1967), Folk (1976), Walker (1979), Breed and Breed (1979) and El Baz (1978), that dunes become redder with distance in the direction of transport and hence as the sand in them becomes older. However, Gardner and Pye (1981) suggest that colour is

48

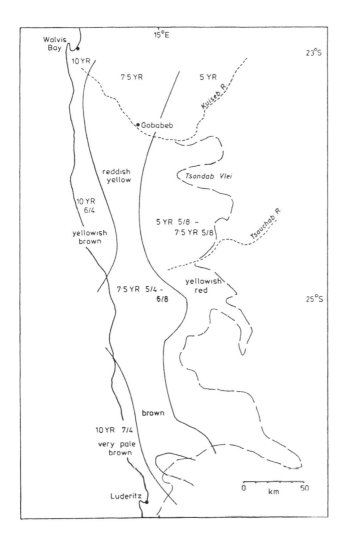

Figure 22. Distribution of sand colours in the Namib Sand Sea.

not necessarily a function of dune age and that time is not always an independent variable. As observed by Folk (1976), reddening is a product of moisture availability, temperature and time. Also important are grain mineralogy, presence of weatherable minerals and the aeolian dust input. Further, different sources for sands, especially the existence of pre-reddened sediments, may be important (Wasson 1986).

The spatial variability of dune sand colours in the Namib Sand Sea (Fig. 22) suggests that colour does not vary with transport distance and age, as suggested by Logan (1960). If this was the case, then linear dunes would become redder along their length from south to north. There is no evidence for this, and individual dunes have essentially the same colours throughout their length. The strongest correlations observed are between sand colour and climatic parameters. Colours become redder as rainfall and temperature increase from west to east across the desert. Thus, increased reddening is a function of higher temperatures and moisture availability. The role of fog precipitation as a moisture source is uncertain. However, there are some sharp discontinuities in dune colour as dune type changes, especially between pale crescentic dunes and redder linear dunes in the western parts of the sand sea. Elsewhere, as at site XXVa, star dunes are much redder than nearby linear dunes. In the eastern parts of the sand sea, relatively well vegetated compound linear dunes are redder than the barer, more active complex linear dunes to the west. These observations tend to support Gardner and Pye (1981) who argued that increased sand movement and more active dunes give rise to paler colours through abrasion of coatings on grains and decreased opportunities for weathering.

Colour variations may also reflect different sediment sources for the sand. Thus dunes in the eastern Namib may be red because they are derived in part from nearby outcrops of the Tsondab Sandstone Formation, as suggested by Besler (1980). Coastal crescentic dunes may be pale as they are derived from pale coloured beach sands.

Dune colour in the sand sea is therefore a complex function of sediment source, dune activity and regional changes in temperature and moisture availability. The pale colours of the coastal dunes probably result from their relatively recent derivation from coastal sediment sources, together with the hyper-arid climate of this zone and the active aeolian environment. Reddening inland is the result of higher temperatures and humidity, coupled with probably less active sand movement. The red dunes of the eastern margins of the sand sea reflect their relative stability, distance from source zones and the less arid climate of this area. Locally, derivation of dunes from red sands of the Tsondab Sandstone Formation may be important.

4.2 GRAIN MORPHOLOGY

Although early workers (e.g. Shotton 1937; Cailleux 1952) suggested that aeolian sands were rounded or well rounded in shape, modern investigations (Folk 1978; Goudie and Watson 1981) indicate that, in aeolian sands, true roundness in the dominant 2.0-3.0 phi size group is rare and most grains are subangular to subrounded in shape. Goudie and Watson (1981) also noted that grains from different sand seas cluster around distinctive grain roundness characteristics.

As observed by Besler (1980), a variety of grain morphologies occur in the Namib Sand Sea, with both frosted and clear quartz occurring. Clear quartz grains, usually subangular to subrounded in shape, are particularly common. Frosted grains tend to be larger and more rounded and make up a higher proportion of the samples from eastern parts of the sand sea. Conchoidal fracturing and crescentic impact scars on clear quartz grains were noted in some samples, especially from the western part of the sand sea. This may reflect salt weathering of grains in the manner suggested by Goudie et al. (1979), or impacts generated by the high energy wind regime of this area (Lancaster 1985a).

In the Namib Sand Sea, sand grains in the 2.0-3.0 phi size range (i.e. the dominant size group in most areas) vary in roundness between subangular to rounded, with most grains being subangular or subrounded. The mean grain roundness on Folk's logarithmic transformation of Power's roundness scale (Folk 1978) is 3.36, or subrounded, with a standard deviation of 0.72. This is very similar to the values obtained from the Namib by Goudie and Watson (1981). Grains coarser than 2.0 phi tend to be rounded or well rounded.

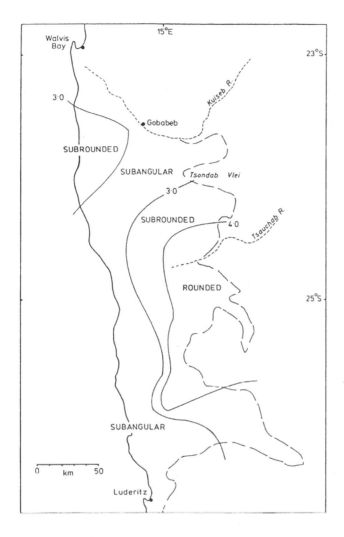

Figure 23. Variation in grain
roundness in the Namib Sand
Sea, as measured by Folk's
logarithmic transformation of
Power's roundness scale.

The pattern of grain roundness in the Namib Sand Sea (Fig. 23) shows that in the 2.0-3.0 phi range grains tend to be subangular in shape in western, northern and some southern parts of the sand sea and rounded to well rounded in eastern areas. This pattern may reflect transport and rounding in the direction of transport from west to east, or possibly sources of sand in pre-existing aeolian sandstones of the Tsondab Sandstone Formation.

Folk (1978) has suggested that there may be a correlation between dune type and the degree of rounding of constituent sands. In the Namib Sand Sea there is little support for this and sand from all dune types ranges between subangular and subrounded.

As observed by Goudie and Watson (1981) there appears to be little change in grain roundness between basal and crestal areas of dunes of linear and star form. However, where plinth sands are particularly coarse the proportion of rounded grains in the sample tends to increase.

4.3 GRAIN MINERALOGY

4.3.1 *Light minerals*

The sands of the Namib Sand Sea are composed dominantly of quartz. Feldspar constitutes approximately 10% of the light minerals in most samples, and is a similar shape and size as the quartz grains.

4.3.2 *Heavy minerals*

The composition of the heavy minerals in the dune sands is summarised in Appendix 1. The sands are dominated by clinopyroxene, garnet and opaque minerals (Lancaster and Ollier 1983). Other minerals are present as accessory or rare minerals. Clinopyroxene is the most common transparent mineral and occurs in well rounded elongate grains. It varies in colour from clear through yellow and pale green to pale brown. Pink garnets are the second most common heavy mineral and often are present as large 2.0-2.5 phi grains. Opaque minerals are dominated by magnetite and illmenite. Also present in varying quantities are hornblende, epidote, staurolite, rutile and zircon. Slightly abraded biotite flakes are present in some samples, especially those from sites III, XXI, XXIII and XXV. In the case of sites XXI and XXIII derivation from nearby granite outcrops seems likely as biotite is unlikely to survive much aeolian transport. In the case of site III, derivation from Kuiseb River fluvial sediments and input by N-NE winds seems probable.

Essentially the same suite of minerals is found throughout the sand sea and indicates derivation of the sands from metamorphic rocks which are relatively rich in quartz, feldspar, garnet and clinopyroxene. There is relatively little zircon or rutile in the assemblage. Significantly, the content of weatherable minerals is quite high, indicating that the sands are still relatively 'immature'.

According to Lancaster and Ollier (1983) two main varieties of the heavy mineral suite can be distinguished: one dominated by clinopyroxene and the other dominated by large grains of garnet. The distribution of the two suites is shown in Figure 24. Although there is no clear cut pattern, the garnet rich suite dominates in eastern and north eastern areas of the sand sea. These variations may possibly reflect two sources for the sands of the sand sea: one coastal and one inland.

4.4 GRAIN SIZE AND SORTING CHARACTERISTICS OF DUNE AND INTER-DUNE SANDS

Sands were sampled at 26 major sites throughout the sand sea (Fig. 21), to include most of the major dune types identified in Chapter 3 (Fig. 25).

In the literature on dune sediments, it is frequently very difficult to identify the place on a dune to which descriptions of sand characteristics refer, as the location of samples is frequently not given. In view of the spatial variability of the grain size and sorting parameters of dune sands (Folk 1971) this makes detailed comparisons between the sands of different sand seas and dunefields difficult or impossible. Further, for comparisons to

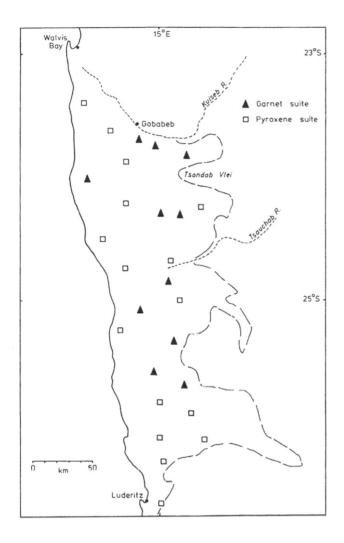

Figure 24. Distribution of heavy mineral suites in the Namib Sand Sea (after Lancaster and Ollier 1983).

be made with the rock record, it is important to know which stratification type is being sampled. Consequently, an effort was made to make the location of sampling points as consistent and systematic as possible, by adopting standard sampling schemes for each dune type. Samples were taken from the dune crest and the mid point of each of the morphologic and process sub environments identified in Figure 26. At each site, samples of surface sand were collected by scraping together the top 10 mm of sand over an area of approximately 0.50 m², to yield an average sample weight of 500 gm. Thus, most sediments sampled represent those of wind ripples. On slip faces most deposits sampled were from grainflow strata. In total some 1500 sediment samples were collected and analysed.

Following determination of sediment colour using a Munsell Soil Colour Chart, samples were quartered and seived through a set of seives at half phi intervals. Graphical

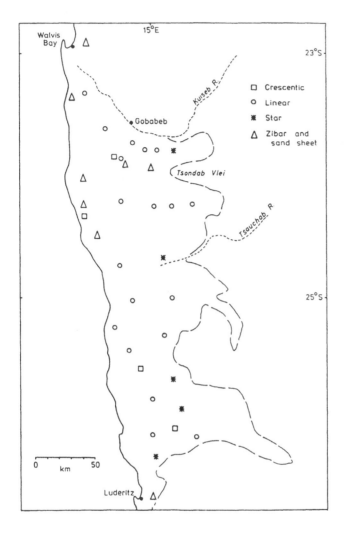

Figure 25. Dune types at sediment sampling locations.

measures of grain size and sorting were computed following the methods of Folk and
Ward (1957). Mean values of grain size and sorting parameters and textural characteristics for each site, dune type and sub environment are tabulated in Appendices 2 and 3.

4.4.1 Crescentic and barchan dunes

Sediments of crescentic and barchan dunes were sampled at eight sites in the sand sea
(Fig. 25). Typical cumulative curves and size-frequency histograms for each dune sub
environment are given in Figure 27. The grain size and sorting characteristics of these
dune types (Table 5) appears to be much more variable than that of other dune types,
probably as a result of different sources of sand for each group of dunes and their close
proximity to source zones.

54

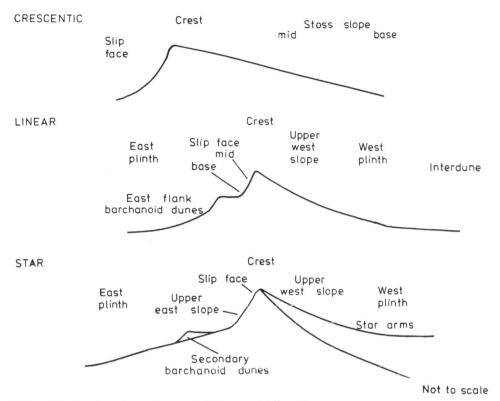

CRESCENTIC

Crest

Stoss slope

Slip face

mid

base

LINEAR

Crest

East plinth

Slip face

Upper west slope

West plinth

mid

base

Interdune

East flank barchanoid dunes

STAR

Crest

East plinth

Upper east slope

Slip face

Upper west slope

West plinth

Star arms

Secondary barchanoid dunes

Not to scale

Figure 26. Location of sampling points for dunes of different types

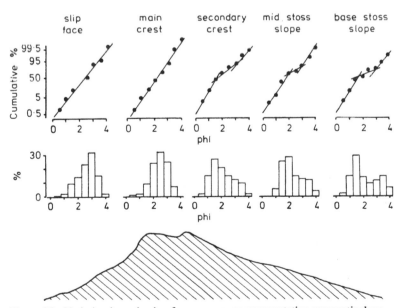

Figure 27. Variation in grain size-frequency over representative crescentic dune.

55

Table 5. Mean values of grain size and sorting parameters for crescentic and barchan dunes (phi units)

Position	Mean	Standard deviation	Skewness	Kurtosis
Crest	2.20	0.55	0.19	0.50
Slip face	2.32	0.54	0.07	0.49
Mid stoss slope	2.24	0.73	0.16	0.47
Base stoss slope	2.10	0.84	0.26	0.48

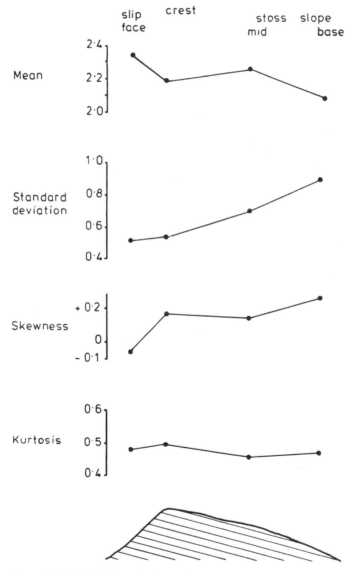

Figure 28. Variation in grain size and sorting parameters over a representative crescentic dune.

The mean grain size of crescentic and barchan dune sands exhibits a wide variation from one sampling site to another. Crest sands from the majority of crescentic and barchan dunes vary between 2.00 and 2.30 phi (0.25 to 0.20 mm). This is rather coarser than most linear or star dunes. Similar degrees of variability exist for other sub environments, but despite this, there is a consistent pattern of grain size variation over the dunes (Fig. 28). Sand becomes finer from the base of the stoss slope towards the crest and slip face, with average values of mean grain size decreasing from 2.10 phi (0.22 mm) at the base of the stoss slope to 2.32 phi (0.20 mm) on the slip face.

Sorting, as measured by phi standard deviation (σ_I), exhibits much less variability from site to site than mean grain size. Values of σ_I for most crescentic and barchan crest sands range between 0.40 and 0.52, or well to moderately well sorted. Sorting of crest sands of these dunes is generally poorer than that from the crests of linear or star dunes. On each dune sorting improves from 0.70-0.85 (moderately sorted) on the lower parts of the stoss slope to 0.45-0.55 (well sorted to moderately well sorted) at the crest and on the slip face.

Sands from crescentic and barchan dunes are either strongly positively skewed with phi skewness values between 0.15 and 0.35 or near symmetrical to slightly negatively skewed (phi skewness +0.10 to -0.10). Sands from the base of the stoss slope are quite strongly positively skewed (phi skewness 0.26 to 0.52) due to an abundance of coarse grains and a tail of fines. Skewness decreases up the stoss slope to the crest and slip face. For example, at site XI, skewness values decline from 0.10 at the base of the stoss slope to 0.07 at the crest and become negative (-0.01 to -0.05) on the slip face. Elsewhere, especially where the overall mean grain size is coarse, skewness changes from strongly positively skewed to less positively skewed over the dune.

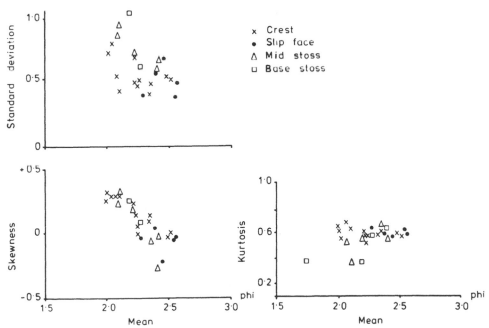

Figure 29. Relationships between site mean values of grain size and sorting parameters in areas of crescentic dunes.

Values of phi transformed kurtosis for crescentic and barchan dune sands range between 0.43 to 0.54. Stoss slopes tend to be mesokurtic and crests and slip faces leptokurtic.

Bivariate plots showing the interrelationships between site mean values of mean grain size, standard deviation, skewness and kurtosis are presented in Figure 29. It was not possible to pick out distinct groups of sands which correspond to specific sub environments on the dune and the changes in grain size and sorting parameters appear to be continuous, rather than discrete.

There are distinct trends in the relationships between mean grain size, standard deviation and skewness. As sands from crescentic dunes and barchans become finer, sorting improves and skewness becomes less positive, even becoming negative in some instances, especially on slip faces. Similar relationships have been noted for crescentic dunes in a world wide sample (Ahlbrandt 1979) and from detailed studies of dunes in the Skeleton Coast dunefield, Namibia (Lancaster 1982a); Saudi Arabia (Binda 1983, Vincent 1984) and the Great Sand Dunes, Colorado (McKee 1983). A clear association between coarse sands and strong positive skewness was noted by Folk (1971).

4.4.2 *Linear dunes*

Linear dune sediments were investigated at 17 sites in the sand sea (Fig. 25), with samples taken from the sub environments indicated in Figure 26. Table 6 shows mean values of grain size and sorting parameters for compound and complex linear dunes, whilst typical cumulative curves and size frequency histograms are given in Figures 30 and 31.

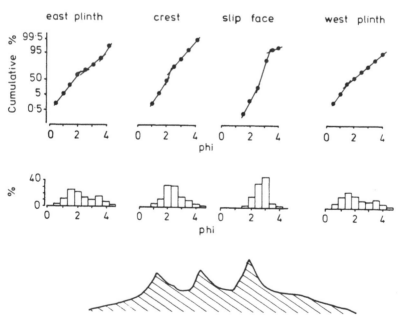

Figure 30. Variation in grain size-frequency across a representative compound linear dune.

Table 6. Mean values of grain size and sorting parameters for linear dunes (phi units).

Position	Mean	Standard deviation	Skewness	Kurtosis
Compound linear dunes				
Crest	2.25	0.39	0.19	0.50
Slip face	2.33	0.41	0.04	0.47
Plinth	2.01	0.86	0.22	0.46
Interdune area	1.96	1.04	0.29	0.46
Complex linear dunes				
Crest	2.49	0.36	0.13	0.51
Slip face	2.51	0.37	0.03	0.50
Upper west slope	2.40	0.51	0.09	0.49
Plinth	2.07	0.76	0.32	0.48
Interdune area	1.98	0.90	0.34	0.47
E flank barchanoid dunes	2.30	0.44	0.17	0.52
Corridor crossing dunes	2.19	0.48	0.26	0.55

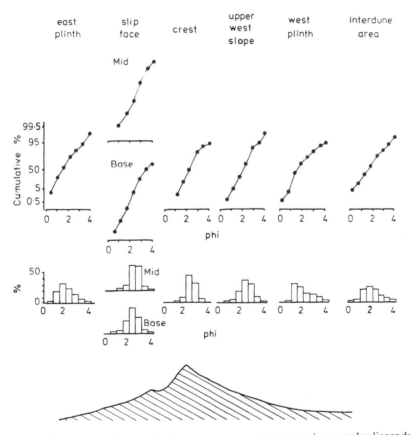

Figure 31. Variation in grain size-frequency across a representative complex linear dune.

Although regional changes in the grain size and sorting character of linear dune sands give rise to a wide spread in values for each of the parameters, there is a consistent pattern of changes in grain size and sorting at all sites (Fig. 32). Thus, sand from crests, slip faces and upper windward slopes is consistently finer and better sorted and more symmetrical than that from plinths and interdune areas. This was recognised by Goudie (1970) and Besler (1976) and confirmed for the northern part of the sand sea by Lancaster (1981a) and for the whole sand sea by Lancaster (1983b). These grain size patterns are similar to those observed by Bagnold (1941) from the Western Desert, Egypt; by Alimen (1953) in Algeria; McKee and Tibbitts (1964) in Libya; Glennie (1970) in Oman; Tsoar (1978) and Sneh and Weisbrod (1983) in Sinai; and Lancaster (1986) for some linear dunes in the south western Kalahari.

Linear dunes in the Namib Sand Sea are composed of fine to medium sand with a mean grain size varying from 1.80 to 2.75 phi (0.27 to 0.15 mm). There is a steady decrease in

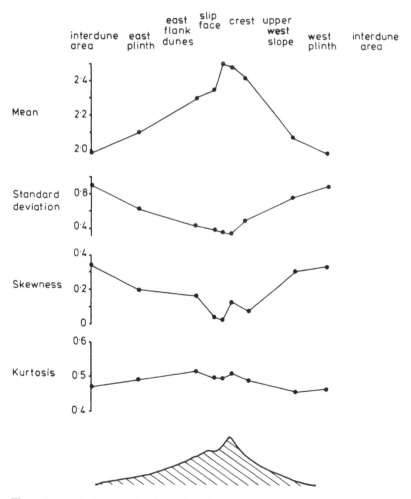

Figure 32. Variation in grain size and sorting parameters across a representative complex linear dune.

mean grain size from interdunes to the crest and slip face. On complex linear dunes, mean grain size decreases from 1.96-1.98 phi (0.25-0.26 mm) in interdunes through 2.01-2.07 phi (0.24-0.25 mm) on the plinths to 2.35-2.45 phi (0.18-0.20 mm) on upper windward slopes and 2.49 phi (0.18 mm) on the crests. Sand is still finer, at 2.51 phi in the mid slip face position, but coarser (mgs 2.32 phi, 0.22 mm) at the base of the slip face. Often, western plinth sands tend to be slightly coarser than those from the eastern plinth. Sand from crests of superimposed barchanoid dunes on the east flank of complex linear dunes is very similar (mgs 2.44 phi) to that of the upper parts of the dune with which they are linked.

In areas of compound linear dunes, values of mean grain size of interdune and plinth sands are similar to those from equivalent sub environments in areas of complex linear dunes, but sand from upper western slopes, crests and slip faces tends to be coarser than that from equivalent locations on complex linear dunes. Thus, the mean grain size of the crests of compound linear dunes averages 2.34 phi (0.21 mm) whilst that from slip faces averages 2.33 phi (0.22 mm).

Sorting (σ_I) varies over the dunes in a similar way to mean grain size (Fig. 32). Interdunes and plinths are moderately to poorly sorted ($\sigma_I = 0.76$-1.04), whilst upper

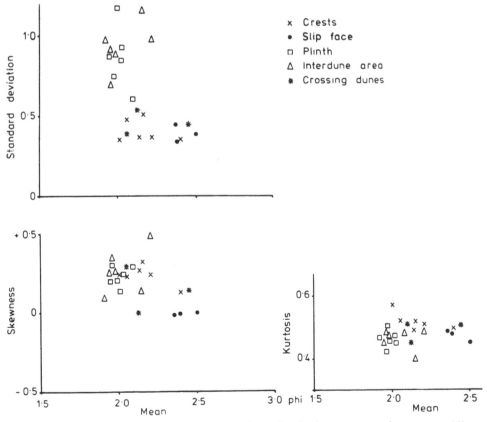

Figure 33. Relationships between the site mean values of grain size parameters for compound linear dunes.

windward slopes tend to be moderately well sorted (0.51). The best sorted sands are found at the crest or on the mid slip face ($\sigma_I = 0.36\text{-}0.41$; very well to well sorted) and on the crests of east flank barchanoid dunes. Sand from all sub environments on compound linear dunes tends to be slightly less well sorted than that from complex linear dunes.

Most linear dune sands are positively or fine skewed. Those from interdunes and plinths are strongly fine skewed with mean phi skewness values of 0.22-0.29. Crest sands

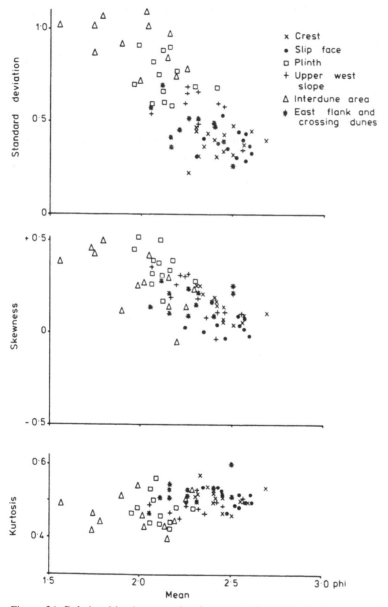

Figure 34. Relationships between the site mean values of grain size parameters for complex linear dunes.

are slightly positively skewed or near symmetrical (0.09-0.19) whilst slip face sands are near symmetrical (0.03-0.04) and occasionally slightly negatively skewed.

Transformed kurtosis values (kg_l) show that many plinth and interdune sands are platykurtic with mean transformed kurtosis values of 0.46, indicating less active sorting, compared to crest and slip face sands which are mesokurtic (0.50-0.51).

Bivariate plots of average values of mean grain size, sorting, skewness and kurtosis for each site (Figs. 33 and 34) show that, as mean grain size decreases, sorting improves and skewness decreases. It is also clear that, for each site, sand from crests, slip faces and upper western slopes is consistently finer, better sorted and more symmetrical than that from plinths and interdune areas.

Linear dunes are thus composed of two groups of sands: a fine, well sorted near symmetrical to slightly positively skewed crest, slip face and upper western slope group, and a coarser, moderately to poorly sorted plinth and interdune group. Sands from east flank superimposed dunes are very similar in composition to those from the crest and slip face.

A linear discriminant analysis was used by Lancaster (1981a) to establish that these two populations of sands are statistically significantly distinct. Application of similar techniques to linear dune sands from all sites sampled confirms this hypothesis. Although there is some overlap between the groups, they may be considered statistically significantly different, with standard deviation being the parameter which most clearly discriminates between groups, followed by mean grain size. However, recent work by Watson (1986) and Livingstone (1986a) based upon closely spaced sampling of linear dunes indicates that grain size changes across Namib linear dunes are continuous and discrete groups of sands cannot be identified. The two groups of sands identified by discriminant anlysis therefore represent the end members of a continuum of grain size and sorting changes across the dunes.

4.4.3 *Star dunes*

Star dune sands were studied at five sites in the sand sea (Fig. 25). Samples were taken from up to eight sub environments (Fig. 26). Mean values of grain size and sorting parameters are summarised in Table 7. Typical cumulative curves and grain size frequency histograms for each sub environment are shown in Figure 35.

Star dunes in the Namib Sand Sea are composed of fine to very fine sands with a range

Table 7. Mean values of grain size and sorting parameters for star dunes (phi units).

Position	Mean	Standard deviation	Skewness	Kurtosis
Crest	2.29	0.29	0.13	0.53
Slip face	2.46	0.30	0.12	0.52
Upper slopes	2.34	0.35	0.16	0.53
Plinth	2.17	0.54	0.16	0.55
Interdune	2.04	0.81	0.11	0.51
E flank barchanoid	2.30	0.31	0.17	0.51

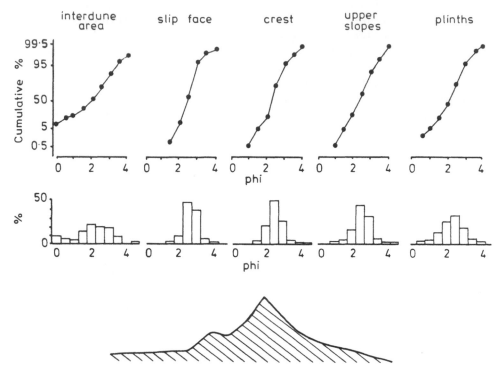

Figure 35. Variation in grain size-frequency across a representative star dune.

in mean grain size between 2.04 and 2.74 phi (0.24 to 0.15 mm). As with linear dunes, there is a decrease in mean grain size from interdunes to crests and slip faces (Fig. 36). Mean grain size decreases from an average of 2.01 phi (0.25 mm) in interdune areas to 2.17 phi (0.22 mm) on plinths and 2.29 phi (0.22 mm) at the crest of the dune. Slip faces and the upper windward slopes of the dunes are finer still, with mean grain sizes of 2.46 phi (0.18 mm) and 2.34 phi (0.20 mm) respectively. Most star dunes have small dunes, often of barchanoid or crescentic form, superimposed on their mid and lower slopes. Sand from the crests of these dunes is essentially similar to that from the crest of the main dunes, with an average mean grain size of 2.30 phi (0.22 mm).

Sorting of star dune sands, is typically moderate to very good. Interdune areas tend to be moderately sorted with an average σI value of 0.81. In some areas very coarse poorly sorted interdune sands occur in proximity to local rock outcrops (e.g. at site XIV) or fluvial sediments (e.g. at site X). Plinths are moderately well to very well sorted, with values ranging from 0.87-0.38 with a mean of 0.54. Sand from the crests, slip faces and upper windward slopes of star dunes is always very well to well sorted with σI values averaging 0.29-0.35.

Sands from star dunes are generally positively skewed. Skewness values are highest in interdunes and plinths with a mean value of 0.16-0.23 (strongly fine skewed) and decrease towards the crest and upper slopes of the dunes, where average phi skewness values range from 0.07-0.15, or near symmetrical to slightly fine skewed.

Phi transformed kurtosis values for star dune sands tend to be somewhat higher than for other dune types. Mean kurtosis values range from 0.51-0.55, or slightly leptokurtic, which indicates that sorting processes are very active on these dunes. As on linear dunes, the variability in kurtosis values is not very large, but there is a tendency for lower kurtosis values to be observed from plinth and interdune areas.

Bivariate plots (Fig. 37) of the site mean values of grain size and sorting parameters show that, as with linear dunes, the crests, upper slopes and slip faces of star dunes are consistently finer and better sorted than adjacent plinths and interdune areas. Considering

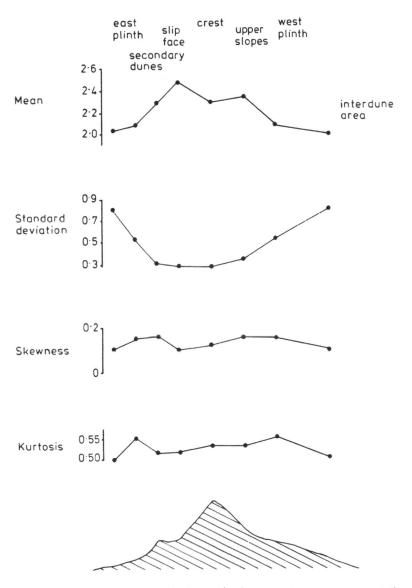

Figure 36. Variation in mean grain size and sorting parameters over a representative star dune.

all sites together, there is little difference in skewness values between crestal and lower areas of the dunes, unlike the situation on linear dunes. However, when each site is considered individually, consistent differences do occur and plinths and interdunes are more positively or fine skewed than upper areas of the dune. Thus, sands of star dunes become finer and more symmetrical and more leptokurtic as sorting improves. As with linear dunes, the sands of star dunes can be considered to fall into two distinct groups, a basal or plinth group and an upper or crestal and slip face group. Sorting, as measured by phi standard deviations, is the variable which is most important in discriminating between groups.

4.4.4 *Zibar and sand sheets*

In many areas of linear and star dunes, the interdune areas are composed of gently undulating medium to coarse sands. Locally, the undulations reach an amplitude of 1-2 m and are sufficiently prominent to enable bedforms of the type known as zibar (Holm 1960; Warren 1972) to be recognised. South of the Uri Hauchab mountains at site XVI, undulating sands described as 'low rolling dunes without slip faces' by Lancaster (1983a) are a very well developed form of zibar with an amplitude of 10-20 m. In some parts of

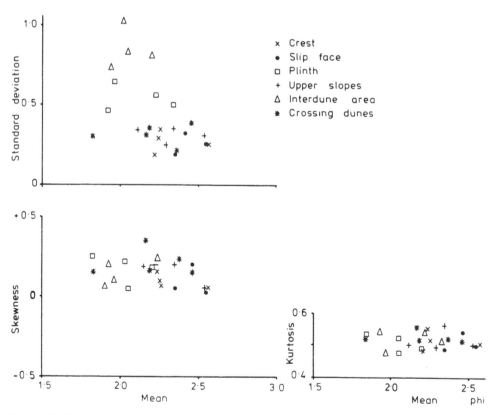

Figure 37. Relationships between site mean values of grain size and sorting parameters for star dunes.

the southern part of the sand sea (e.g. site XXIV), there are extensive areas of coarse sand without bedforms, termed sand sheets.

Both zibar and sand sheets are considered together, in that both are characterised by coarse, often bimodal, sands, which are generally poorly sorted. The size frequency and grain size and sorting characteristics of zibar and sand sheets are summarised in Figure 38 and Table 8 respectively.

The mean grain size of zibar and sand sheets is coarser than most other dune types, except some transverse dunes. It ranges from 2.23 to 1.84 phi (0.21 to 0.28 mm) with an average value of 2.05 phi (0.24 mm). Where zibar are large enough, some variability in mean grain size from troughs to crests is evident. At site XVI, mean grain size decreases from 1.83 phi (0.28 mm) in troughs to 1.99 phi (0.25 mm) on the crests.

Sand sheet deposits tend to have very similar grain size characteristics to zibar with average mean grain size values of 2.02 phi (0.24 mm). Many zibar and sand sheet deposits are bimodal with a coarse mode centering on 1.5 or 2.0 phi and a fine mode at 3.0 or 3.5 phi.

Most zibar and sand sheet deposits are moderately or poorly sorted, with σ_I values ranging between 0.7 and 1.10, with a mean of 0.87. Sand sheets are poorly sorted, with phi standard deviations averaging 1.03.

Skewness values of zibar and sand sheet deposits are widely scattered, and range from strongly fine skewed (phi skewness 0.56) to near symmetrical to slightly negatively skewed (-0.15). Most samples cluster between 0.45 and 0.07 with a phi mean skewness of 0.26, or distinctly fine skewed.

Values of phi transformed kurtosis exhibit less variability than do other parameters, ranging between 0.40 and 0.49 with a mean of 0.46. All are therefore classified as platykurtic, with best sorting in the tails of the distribution.

Table 8. Mean values of grain size and sorting values for zibar and sand sheets (phi units).

	Mean	Standard deviation	Skewness	Kurtosis
Zibar	2.05	0.87	0.26	0.46
Sand sheets	2.02	1.03	0.22	0.45

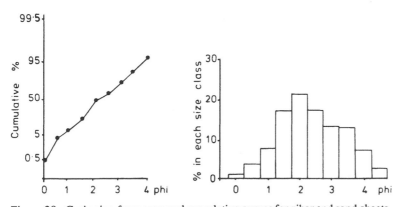

Figure 38. Grain size-frequency and cumulative curves for zibar and sand sheets.

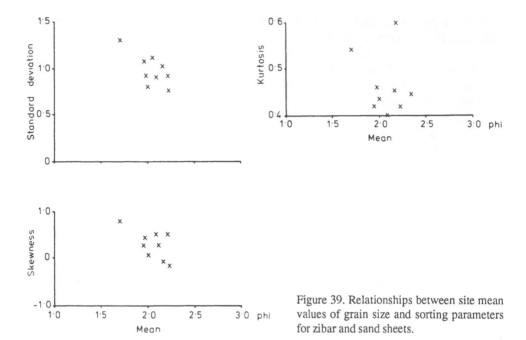

Figure 39. Relationships between site mean values of grain size and sorting parameters for zibar and sand sheets.

Bivariate plots of grain size and sorting parameters for zibar and sand sheets (Fig. 39) show very similar patterns and trends as those for other dune types. Improved sorting and lower skewness values are associated with a fining trend and smaller mean grain sizes. As with other dune types, there is little variation in kurtosis values with grain size.

4.4.5 *Comparisons between the grain size and sorting character of dunes of different types*

Figure 40 shows bivariate plots of the relationships between site mean values of grain size and sorting parameters for the crestal sands of dunes of different types. Several general relationships between dune types and sand characteristics are evident. There is an improvement in sorting, overall decrease in mean grain size and slight increase in the kurtosis values of crest sands from zibar and sand sheets, through barchans and crescentic dunes to linear and star dunes, which tend to be best sorted, although not always the finest. There is a wide spread of values from crescentic dunes, some of which, for example at sites IX, XXI and XXII, are clearly coarser and less well sorted than most linear dune sands. However, crescentic dune crests at site XXIII are very similar to those of compound linear dunes elsewhere in the southern parts of the sand sea. Sands from compound crescentic dunes at site XI have a similar mean grain size to, but are much less well sorted than adjacent linear dunes at site III.

The clearest relationship to emerge is that zibar and sand sheets are composed of coarse, poorly sorted sands which have strong positive skewness and low transformed kurtosis values. Similar relationships have been observed in Saharan sand seas by Capot

Rey (1947), McKee and Tibbitts (1964) Warren (1972) and Maxwell (1982); in Sinai (Tsoar 1978); the Skeleton Coast dunefield (Lancaster 1982a) and the Algodones dunes, California (Kocurek and Nielson 1986 and Nielson and Kocurek 1986).

It is not clear whether the observed relationships between grain size and sorting characteristics and dune types are genetic. Apart from the association between zibar and coarse, poorly sorted sands, few consistent relationships seem to occur. In the Algerian Sahara, Bellair (1953) found that barchans and crescentic dunes were composed of well sorted, unimodal sands, but complex linear and star dunes were composed of bi or tri-modal sands. However, Alimen et al. (1958) and Capot Rey and Gremion (1964) could find no consistent relationships. At White Sands, New Mexico, McKee (1966) observed a progressive decrease in grain size and improvement in sorting from dome dunes, through barchans and crescentic dunes to parabolic dunes. In the Sudan, Warren (1970) demonstrated that sands from undulating sand sheets were coarser and less well sorted than those from crescentic dunes, which were in turn coarser and less well sorted than adjacent linear dunes. Many of the relationships can be best explained by progressive sorting and fining of sands downwind from sediment sources, as at White Sands, or by different sources of sand for different dunes. In the Namib Sand Sea, the latter argument can be employed to explain differences between adjacent dune types at Sossus Vlei, where small barchanoid and reversing dunes on the vlei surface are much coarser and less well sorted (average mean grain size 2.06 phi; 0.24 mm; $\sigma_I = 0.49$) than adjacent star dunes (mean grain size 2.38 phi; 0.19 mm; $\sigma_I = 0.27$). Differences in colour between the grey brown barchans and reddish yellow star dunes clearly indicate a different sediment sources for each dune type.

North west of the Tsondab Flats (site IX), the north-south trending linear dunes are discontinuous where they cross former drainageways. Gaps in the linear dunes are filled by crescentic and barchanoid dunes aligned transverse to south westerly winds and low simple linear dunes on WSW-ENE alignments. Between the dunes are extensive areas of zibar and sand sheets. Figure 41 shows how mean values of grain size and sorting vary between dune types in this area. Thus zibar are much coarser and less well sorted than the crest sands of adjacent crescentic ridges which have average mean grain sizes in the range 2.31-2.36 phi (0.19-0.20 mm) and average phi standard deviation values (σ_I) between 0.55-0.60. The crest sands of nearby simple linear dunes on WSW-ENE alignments are finer still and much better sorted (mean grain size 2.46 phi; 0.18 mm; $\sigma_I = 0.35$).

The observed differences cannot be explained in terms of different sand sources, for each dune type links to another. Rather, these differences in grain size and sorting, as with many of those observed in the Namib Sand Sea, can best be explained in terms of the pattern of sand movements on dunes of different types, as originally suggested by Folk (1971). On crescentic dunes, sand movement is essentially unidirectional, and saltating sands are buried on avalanche faces to be recycled later as these deposits are exposed by advance of the dune. Crestal areas of linear dunes are reversed seasonally and undergo more frequent resorting. Star dune crests are best sorted, because they undergo constant resorting by multi directional winds.

However, on the scale of the sand sea, differences in grain size and sorting parameters which are related to sediment movement patterns on dunes of different types may be subsumed by large scale differences related to the position of the dune relative to sand transport paths. Thus barchan and crescentic dunes are often found in upwind near source

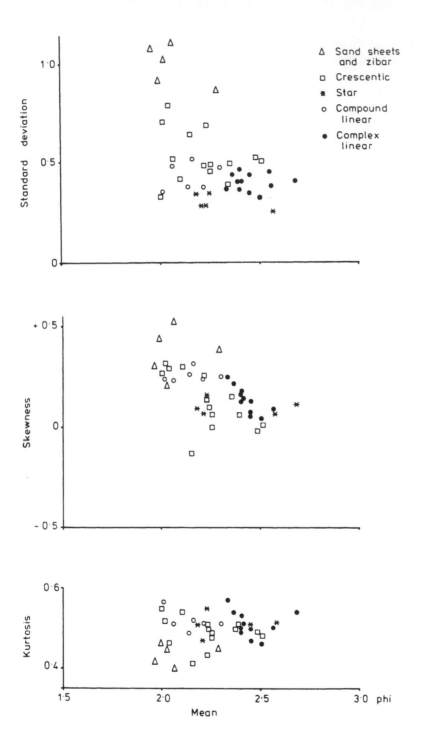

Figure 40. Relationships between site mean values of grain size and sorting parameters for dunes of different types.

70

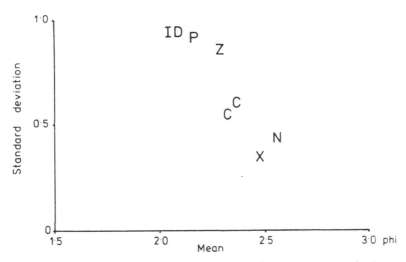

Figure 41. Relationships between mean grain size and sorting parameters for dunes of different types at site IX. ID: interdune area between linear dunes. Z: zibar, C: crests of crescentic dunes; X: crests of WSW-ENE trending simple linear dunes, N: crests of north-south trending linear dunes; P: plinths of north-south trending complex linear dunes.

areas, where coarser sands may be more common and star dunes in downwind accumulation zones dominated by finer sands.

4.4.6 *Comparisons with other sand seas and dunefields*

Despite differences in the character of source sediments and aeolian environments, considerations of the process of sand movement by the wind suggest that all aeolian sands should evolve towards similar grain size and especially sorting characteristics (Bagnold 1941). Ahlbrandt (1979) assembled a variety of data on grain size and sorting character of inland dune sands and noted the wide range of values for grain size and sorting parameters. Although there is a considerable amount of literature on the sediments of desert dunes, comparisons of grain size and sorting parameters between different sand seas and dunefields are hampered by the lack of data which refers to specific dune types and positions on dunes. For this reason, data for dune crests only are compared. Further, only a proportion of studies use widely accepted (e.g. Folk and Ward 1957) type statistics; and many do not report data on skewness and kurtosis values.

Comparative data for average values of mean grain size and phi standard deviation for different dune types in various sand seas are presented in Table 9. Generally, dune sands from the Namib Sand Sea plot towards the finer and better sorted end of Ahlbrandt's (1979) world wide sample of inland dune sands. They lie between the very fine (2.51-2.77 phi; 0.15-0.18 mm) sands of the Thar, Simpson Desert and Gran Desierto sand seas and the somewhat coarser (2.05-2.12 phi; 0.22-0.24 mm) north western Saharan examples of Chavaillon (1964). Compared to other dunefields in southern Africa, dune crest sands from the Namib Sand Sea are generally finer and better sorted than those from the Skeleton Coast and south western Kalahari.

When each dune type is compared, linear dune crest sands from the Namib Sand Sea

Table 9. Grain size and sorting characteristics of sands of different types from different sand seas and dunefields (phi units).

Dune location	Mean	Standard deviation
Linear dunes		
SW Kalahari[5]	2.16	0.49
Libya[1]	2.37	0.46
Canning Basin, Australia[1]	2.02	0.53
Simpson Desert, Australia[6]	2.53	0.43
Thar Desert[7]	2.65	0.56
Sinai[1]	1.87	0.43
Namib Sand Sea	2.44	0.37
Reversing and star dunes		
Great Sand dunes, Colorado[1]	2.09	0.26
Gran Desierto[2]	2.44	0.31
Saudi Arabia[1]	2.67	0.32
Namib Sand Sea	2.29	0.29
Crescentic and barchan dunes		
Gran Desierto[2]	2.43	0.41
Utah[1]	2.39	0.64
White Sands, New Mexico[1]	1.61	0.59
Twentynine Palms, California[1]	2.32	0.48
Arizona[1]	2.32	0.48
Algodones Dunes[8]	2.46	0.42
Tunisia[1]	2.91	0.32
Skeleton Coast, Namibia[4]	2.02	0.51
Namib Sand Sea	2.20	0.55

1. Ahlbrandt (1979); 2. Lancaster et al. (1987); 3. Besler (1980); 4. Lancaster (1982); 5. Lancaster (1986); 6. Folk (1971); 7. Goudie et al. (1973); 8. Lancaster (unpublished data).

fall towards the finer and better sorted end of what is admittedly a small sample. Star dunes in the Namib Sand Sea appear to be somewhat coarser than star dune crest sands elsewhere, but are similarly well to very well sorted. Sand from the crests of crescentic and barchan dunes in the Namib Sand Sea is somewhat coarser and less well sorted than that from comparable dunes elsewhere.

The comparative data exhibit a wide range in mean grain size, but, for each dune type, a surprisingly low range of sorting values as measured by phi standard deviations. Interestingly, they lend some support to the hypothesis that, in general, there is an improvement in sorting from crescentic through linear to star dunes.

Ahlbrandt (1979) concluded that, because their textural character was so variable, it was not surprising that attempts to distinguish aeolian sands from their source material (e.g. Moiola and Weiser 1968) were unsuccessful. However, Ahlbrandt's sample included a wide variety of dune sub environments, which may have given rise to the considerable spread of his data set. Thus, when sands from similar dune sub environments (e.g. the crest) and dune types are compared, it may be possible to identify a more typical 'aeolian' sand.

4.5 SPATIAL VARIATION IN GRAIN SIZE AND SORTING PARAMETERS

Spatial variations in grain size and sorting of dune sands occur at two scales in the Namib Sand Sea: the dune-interdune area unit and the sand sea.

4.5.1 *The dune-interdune area unit*

Spatial variations in grain size and sorting at this scale have been outlined in the descriptions of the grain size and sorting characteristics of each dune type. Common to all dune types is a decrease in mean grain size and an improvement in sorting, as measured by phi standard deviations, together with a decrease in skewness values from interdunes towards dune crest and slip face areas. For linear and star dunes, this pattern can be observed on each side of the dune, which consist of a plinth of relatively coarse, moderately to poorly sorted sand with strong positive skewness, and a crestal and avalanche face zone with finer, well to very well sorted, near symmetrical sands. Usually there is little difference between sands of plinth and interdune areas. On crescentic dunes, mean grain size decreases and sorting improves from lower to upper stoss and crest and slip face areas of the dune. Even on zibar and rolling sand sheet areas of interdunes there is some evidence of a segregation of sands into coarser less well sorted sands in the troughs and finer, better sorted sands in crestal areas. In all cases the dominant pattern is one of fining and improvement of sorting leading to decreasing skewness in the direction or directions of sand transport.

Besler (1976, 1980) attributed differences in grain size and sorting between plinth and crestal areas of linear dunes in the Namib Sand Sea to the presence of 'fluvial' sands in basal areas which were reworked to 'aeolian' sands in crestal areas. She followed a model frequently adopted by many earlier workers in the Sahara (e.g. Alimen et al. 1958; Capot Rey 1970). However, the pattern of grain size and sorting observed on all dune types can best be explained by the operation of the aeolian sorting process. Two models are available to explain the observed patterns. Folk (1971), following Udden (1898), has suggested that sand in the 2.5 phi size group may be selectively transported or winnowed from interdune areas and deposited as dunes. Thus grain size changes effectively represent changes in sorting, with all sands having a similar median grain size, but a varying standard deviation. Thus interdunes are composed of a poorly sorted, bimodal mix of coarse and very fine sands and dunes of well sorted unimodal sand with a mode corresponding to that missing or poorly represented in interdune areas. This model is difficult to apply to dunes in the Namib Sand Sea. Firstly, interdunes generally are not bimodal and secondly they do not appear to be sources for dune sediments.

A second model has been suggested by Lancaster (1981a, 1982b, 1986) in which grain size and sorting differences in aeolian sands can be attributed to variable proportions of a coarse population of sands which is moved by the wind as a traction or creep load and a finer saltating population.

Examination of the size frequency histograms of sands from crescentic dunes (Fig. 27) shows that 40-60% of the sand from the base and middle of the stoss slopes is composed of grains larger than 2.0 phi compared to only 15-20% in crestal and slip face areas. Conversely, 45-55% of sand in these areas falls in the 2.5-3.0 phi classes, yet this size range forms only 15-30% of the sample at the base of the stoss slope. The coarsest grains (larger than 1.0 phi) are usually absent from crest and slip face areas, but form 5-10% of sands at the base of the stoss slope.

Similar patterns occur on linear dunes and star dunes (Figs. 31 and 35), although the grain size-frequency histograms for each group overlap. Sands from the plinth and interdune areas have a mode between 1.0 and 2.0 phi whilst the modal grain size of crestal sands lies between 2.5 and 3.0 phi. Figures 28, 30, 32 and 36 show that, from plinth to crest and slip face areas, the proportions of grains in the coarse fraction decrease steadily whilst the proportion of the finer fraction increases. Thus grains coarser than 1 phi are generally restricted to plinths and interdunes and even the 1-1.5 phi group is rare in crestal areas. Further, the range of grain size fractions occurring in plinth and interdune areas is much greater than in crestal areas. In all cases, these changes in the frequency of different grain size groups lead to a decrease in mean grain size and skewness and an improvement in sorting towards the crest of the dune.

Examination of cumulative probability size-frequency curves for different parts of the dune provides another line of evidence. Using the techniques of Visher (1969), it is possible to identify truncation points where the curves change slope significantly and hence to dissect the samples into sub populations corresponding to those transported by creep, saltation and suspension. In all samples the suspension population is insignificant, which is understandable in that it is normally removed by the wind entirely from the area of a sand sea as aeolian dust. In most samples it is possible to identify a truncation point around 1.5- 2.0 phi, which separates the creep or traction population from the dominant, saltation population. As can be seen from Figure 42, the creep or traction population is strongly represented in the plinth and interdune areas of linear and star dunes and in the basal and mid stoss areas of crescentic dunes. A sample of sand from these areas may be considered to consist of a residual or slowly moving creep or traction population and a passing saltation population. Crestal and slip face sands from all dune types are largely (80-98%) composed of the saltation population.

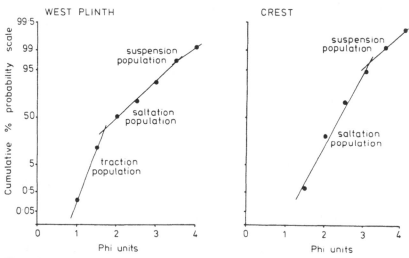

Figure 42. Truncation points on cumulative grain size-frequency curves for plinth and crest sands in areas of linear dunes.

The grain size-frequency distributions of dune sands demonstrate that the main process operating is a loss of the creep or traction population in the direction of transport over a dune. The creep population is therefore concentrated relatively and absolutely in the lower areas of the dune landscape.

In areas of linear and star dunes in the Namib Sand Sea the overall direction of most sand transport is oblique to the trend of the dune and sand crosses from one dune to another. Sand in transport thus consists of a slowly moving group of grains forming the creep or traction load, which most winds are barely competent to move, together with a rapidly moving saltating population. As sand moves across the plinths and up the windward slopes of the dune it encounters areas of steadily increasing slope.

According to Bagnold (1941) most creep or traction load movement takes place by ballistic impact of the saltating sand. The efficiency of this process declines rapidly as slope increases in the direction of transport. The paths of saltating grains change such that angles of impact decrease so that successive saltation of grains ceases when slope angles approach 15° (Rumpel 1985). Grain paths also tend to be diverted along the slope in the manner suggested by Howard et al. (1978), so that probably much of the creep load travels along the dune parallel to its trend. This process is important in developing a fringe of coarse sand around the bases of barchan dunes as observed by Hastenrath (1967) and Warren (1972). The effects of slope are to decrease the rate of creep so that the slower moving traction population is left steadily further behind, resulting in the sands which move towards the crest becoming finer and better sorted.

On reaching the crest the grains are projected beyond the brink in the separation zone, to fall out as grain fall deposits at the top of the avalanche face. Ultimately this zone is oversteepened beyond the angle of repose leading to grain flow, or avalanching in a series of overlapping lobes. True grain fall deposits are thus probably rarely preserved in this situation. In the process of avalanching, the coarser grains appear to become concentrated on the surface and around the edge of the lobes in the manner described by Bagnold (1954). Many roll to the base of the slip face which results in the grain flow deposits becoming finer still. In some crescentic dunes a toe of coarse grains was observed to form at the base of the avalanche face. Many of the very coarse sands found at the base of stoss slopes of crescentic dunes probably result from reworking of this zone by occasional eddies or by the reappearance of basal forset laminae (Finkel 1959; Sharp 1966; Lindsey 1973). In addition, spiral eddies, as described by Tsoar (1983) move sand obliquely along, and even up the slip face and further saltation sorting may take place, as suggested by Sneh and Weisbrod (1983).

Seasonal reversal of winds and sand movement takes place in areas of linear and star dunes in the Namib Sand Sea, so that the processes outlined above operate from two or more directions. Thus both plinths trap creep load whilst coarse grains roll to the base of both east and west facing avalanche faces. The result is that over time, relatively coarse, moderately to poorly sorted sands build up in the plinth and interdune areas, whilst crestal areas of the dunes become finer and better sorted.

4.5.2 *The sand sea*

The spatial variation of grain size and sorting parameters of crest sands over the sand sea is shown in Figure 43 , which follows Lancaster (1982b) and Lancaster and Ollier (1983).

The maps are derived from site mean values of crest sands only. Crest sands were chosen for comparison because they are the only sub environment on the dune common to all dune types.

Figure 43 shows that mean grain size of crest sands tends to decrease from two areas of relatively coarse sands in the southern and western parts of the sand sea. Mean grain size is generally much coarser (2.05-2.24 phi; 0.24-0.21 mm) throughout most of the southern parts of the sand sea and appears to be especially coarse in the area adjacent to

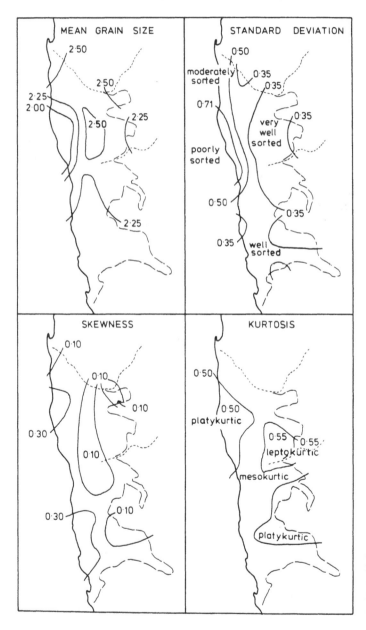

Figure 43. Spatial variation in site mean values of grain size and sorting parameters for dune crest sands in the Namib Sand Sea.

Conception and Meob bays. The finest sands (> 2.50 phi) are to be found in the central and locally in the northern parts of the sand sea. There is a slight tendency for sands to become somewhat coarser along the eastern margin of the sand sea.

Sorting, as measured by phi standard deviation, follows a somewhat similar pattern, with an improvement in sorting from the Conception-Meob area towards the centre of the sand sea and also towards the north, along the belt of crescentic dunes adjacent to the coast. There is a sharp change in sorting values from crescentic to adjacent linear dunes along the inland margin of this zone. Most sands in the southern part of the area are well sorted, becoming very well sorted in central and eastern areas. Areas of very well sorted sands in the southern part of the sand sea are associated with star dune clusters.

The pattern of phi skewness values tends to parallel that for standard deviation and mean grain size, with areas of strongly positively skewed sands in the south west and north west parts of the sand sea and near symmetrical sands in central areas.

Transformed kurtosis values provide an interesting pattern in which platykurtic sands occur in central and western areas and also locally along the south east margins of the sand sea. There is a large area of leptokurtic sands around and north of Sossus Vlei.

The overall pattern of grain size and sorting variation over the sand sea is one of fine, well sorted, near symmetrical sands in central and northern areas of the sand sea and coarser, less well sorted sands in southern and some western parts. There also appears to be an area of coarser, but very well sorted sand, centered on Sossus Vlei.

The patterns are best explained by sand movement away from source zones to the south and west of the sand sea during which coarse grains are left behind in upwind areas. Examination of grain size-frequency histograms for both interdune and crestal sands from linear dunes in different parts of the sand sea (Fig. 44) supports this hypothesis. In the southern parts of the sand sea the modal size group of plinth and interdune sands is 1-1.5 phi, whereas equivalent sands in the northern area have a modal group between 1.5 and 2.0 phi. Similarly, the modal grain size group of the crestal sands of linear dunes in the northern parts of the sand sea lies in the 2.5-3.0 phi groups, compared to a modal size of 2.0-2.5 phi in southern areas.

It appears therefore, that as with the dune-interdune area unit, decreases in grain size

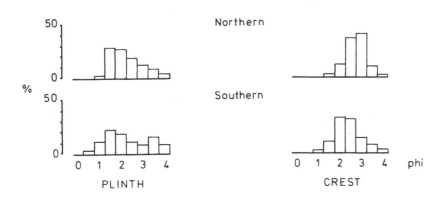

Figure 44. Size frequency histograms for crest and plinth sands in southern and northern parts of the sand sea.

and changes in sorting parameters over the sand sea are achieved by loss of coarse grains in the direction of transport. This has taken place mainly due to different rates of movement of the traction and saltation loads moved by the wind, so that the more slowly moving traction load has tended to remain closer to its original source zones. This is reflected in the southern part of the sand sea and locally in western areas by extensive areas of coarse sands which form a trailing edge or lag deposit of sand sheets and zibar. Similar areas of coarse sands have been noted from the upwind margins of the Gran Desierto Sand Sea and the Algodones dunes by Kocurek and Nielson (1986).

4.6 INTERNAL SEDIMENTARY STRUCTURES OF DUNES

The only systematic investigations of internal sedimentary structures in the Namib Sand Sea are those of McKee (1982) who trenched parts of linear, star and crescentic dunes, together with some interdune areas, in the northern part of the sand sea. In December 1985, heavy rainfall in the Luderitz area enabled investigation of internal structures of linear dunes by the writer. The results of investigations in the Namib appear to confirm more detailed studies of internal structures by McKee and Tibbitts (1964) and McKee (1966, 1982).

4.6.1 *Crescentic dunes*

Structures of crescentic dunes were investigated by McKee (1982) from near Sandwich Harbour (site XI). Trenches at the crest of one of those dunes show cross strata dipping at 33° on the north eastern slope and low angle wind ripple laminae at the crest and on the south western slope. On the leeside of these dunes examples of overturned folds and contorted bedding of laminae were noted. These probably result from overloading during avalanching, aided by wetting of the upper layers of the dune sediments by advective fogs.

Other crescentic dune structures were recorded from dunes crossing the interdune corridors between complex linear dunes south of Gobabeb. Here, avalanche cross strata dip northwards at 24°, with their tops being truncated and covered by a thin veneer of strata dipping south at 3-9°, indicating local wind reversal. Structures in barchanoid dunes near Walvis Bay showed dips towards the north at 30°.

4.6.2 *Linear dunes*

Structures of linear dunes were studied by McKee (1982) from several areas near Gobabeb. In crestal areas of the dunes laminae dip at 28-35° towards the interdune on both sides of the dune crest (Fig. 45A). In the upper parts of the plinths, dip angles are lower (10-20°), declining to 8° or less towards the interdunes. It appears that most of the high angle laminae have developed by avalanching and grain fall generated by both south westerly and north easterly winds. Plinth and interdune laminae mostly represent the deposits of wind ripples.

The crests and western flanks of 15-20 m high SE-NW trending simple linear dunes on the western margins of the sand sea north of Luderitz were trenched by the writer and Ian

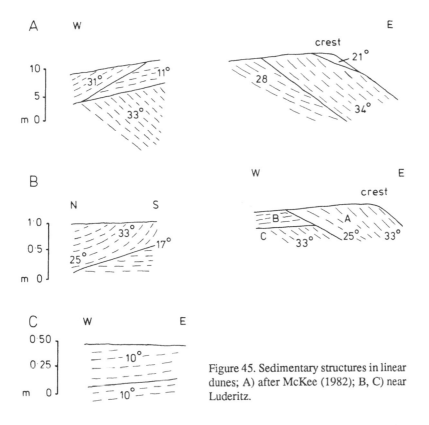

Figure 45. Sedimentary structures in linear dunes; A) after McKee (1982); B, C) near Luderitz.

Corbett in December 1985. The character of the deposits exposed was somewhat different to that observed by McKee (1982) in that cross strata dipping in opposite directions from the dune crests were not observed. Near the dune crest (Fig. 45B) all laminae observed dipped at 25-33° towards the northeast. A prominent bounding surface, at an angle of 25° cuts the laminae deposited by S-SW winds on the currently active slip face and its immediate predecessor. This bounding surface represents erosion of the dune crest and slip face by easterly winds, probably during the winter of 1985. Previous phases of slip face deposition are marked by the laminae of Group C. Overlying them are horizontally bedded wind ripple laminae of Group B, deposited by southerly winds on the crest of a saddle point in the dune.

Deposits on the western slopes of these dunes are of two varieties. Most common are the coarse to finely laminated medium to fine sands with a dip of 5-10° towards the north (Fig. 45C). Towards the dune plinth, dip angles decrease to 2-3° or less and there is an increase of coarse sand, occurring either as 2-3 cm thick laminae or as small lenses.

Towards the crest of the dune northward dipping grainflow and grainfall laminae are frequently preserved, beneath 10-15 cm of horizontal wind ripple laminae (Fig. 45B). Avalanche face laminae dip at angles of 25-33° to the north, pinching out in places and displaying asymptotic curvature indicative of the toe of an avalanche face. In places, a prominent bounding surface dipping NE at 16-18° separates avalanche toe laminae from underlying horizontal laminae. The formation and preservation of avalanche face depos-

W   E

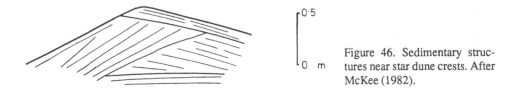

Figure 46. Sedimentary structures near star dune crests. After
McKee (1982).

its in this part of the dune probably results from the migration of peaks and saddles along
the sinuous crestline of the dune in the manner discussed by Tsoar (1982).

In general the internal structures of linear dunes near Luderitz are more similar to those
described by Tsoar (1982) than those investigated by McKee and Tibbitts (1964) in Libya
and McKee (1982) in the northern part of the Namib Sand Sea. It is probable that the
differences observed are a result of variations in wind regime characteristics, as suggested by Lancaster (1982c), following the model of linear dune formation of Tsoar
(1978, 1982). Dunes in areas where formative winds blow at a small angle to the dune, as
in Sinai and the southern Namib, are likely to be characterised by a higher percentage of
wind ripple laminae and less grainfall and grainflow laminae compared with dunes in
areas where winds blow at a high angle to the dune, as in the northern Namib Sand Sea.

4.6.3 *Star dunes*

Structures in the crestal areas of star dunes near Tsondab Vlei and Gobabeb show clear
evidence of seasonal wind reversal. Avalanche face grainfall and grain flow laminae
dipping east at 35° were overlain by another set which covers their eroded tops and dips
towards the west at 33° (Fig. 46). McKee also observed that, in basal areas of star dunes
near Gobabeb, most avalanche face laminae dip steeply to the north east with low angle
strata dipping westward. This pattern was interpreted to mean that the star dune
accumulated largely by winds from south to west and east to south east directions.
Indications of complex wind patterns, due to airflow diversion around the dunes were
also observed by McKee.

5 DUNE PROCESSES

5.1 INTRODUCTION

Following Warren and Knott (1983), aeolian processes can be considered at three spatial and temporal scales. These correspond approximately to the steady, graded and cyclic time scales of Schumm and Lichty (1965). At the steady scale, involving very short or even instantaneous amounts of time and small areas, is the study of the process of sand movement by the wind and the dynamics of small scale bedforms, such as wind ripples. Also probably to be included at this scale of investigations are studies of the primary units of aeolian deposition (Hunter 1977; Schenk 1983). The graded scale involves longer spans of time, in this context ranging from months to tens of years, and particularly concerns the dynamics and morphology of dunes, which may achieve or tend towards an actual or partial equilibrium with respect to rates and directions of sand movements generated by surface winds. Form-flow interactions and feedback processes are important at this scale, which is probably the most important in determining the origins and dynamics of dunes and the accumulation of aeolian sand bodies, yet is the least studied and understood (Lancaster 1984c; Warren 1984). Much of what will be discussed in this chapter concerns the graded scale of process investigations. Study of aeolian processes at the cyclic time scale involves periods of thousands to millions of years and a spatial scale corresponding to that of sand seas and their regional physiographic and tectonic setting. Processes at this scale involve those responsible for the accumulation of the sand sea as a sedimentary body. They will be considered in Chapter 7.

5.2 WINDS AND SAND TRANSPORT PATTERNS

The factors which control the equilibrium morphology of desert dunes are imperfectly understood, but sand supply and wind regime appear to be the most important (Lancaster 1983a; Wasson and Hyde 1983a). In addition, sand grain size and sorting and vegetation cover may be locally important.

Fundamentally, the equilibrium morphology, especially the size, of dunes is a function of their sediment budget, which is the balance between erosion and deposition or sand supply and removal at each point on the dune, summed for the dune as a whole. Further, the size and location of sand seas and dunefields can be regarded as being dependent upon similar principles (Fryberger and Ahlbrandt 1979; Mainguet and Chemin 1983).

In turn, sediment budgets at all scales are dependent upon sand transport rates and thus on the velocity of the wind, and, through the requirements of continuity (Middleton and Southard 1978; Rubin and Hunter 1982), on whether sand transport rates are increasing or decreasing. Consequently, investigation of the patterns of winds and sand transport rates at both the sand sea and dune scales is an important part of the process of understanding the dynamics and morphology of aeolian sand bodies.

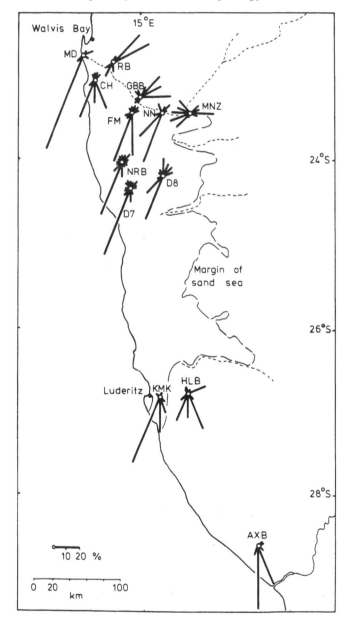

Figure 47. Annual potential sand transport roses for sand sea area.

5.2.1 The pattern of sand movements in the sand sea

The regional pattern of potential sand movements (Fig. 47) reflects comparable patterns in the surface winds, described in Chapter 2, with the effects of the stronger winds emphasised (Lancaster 1985a). In the Namib Sand Sea, two major directional sectors of potential sand movement can be recognised: SSE-SW and E-NE, whilst a third sector, N-NNW, is locally important.

The SSE-SW sector is the most important throughout the sand sea. It is the dominant sand flow sector all the year near the coast where it accounts for 80-90% of sand flow. Inland, the proportion of sand flow from this sector decreases to 55-65% in central areas and 35-40% along the eastern margin of the sand sea. This pattern reflects the decrease in strength and persistence of the sea breeze circulation away from the coast. On the coast, sand flow in this sector is mostly generated by SSE-SSW winds, whilst inland and in the northern and central parts of the sand sea SSW winds are the most important in this sector. In the east of the desert SW-WSW or W winds are the most effective.

The E-NE sector is of minor importance at coastal stations, reflecting the infrequent westward penetration of 'berg' winds, and accounts for less than 10% of annual sand flow. Inland, the importance of this sector increases and it accounts for 10-20% of annual sand flow in central parts of the sand sea and 30-55% on its eastern margins. At stations on the northern edge of the sand sea this sector is dominant, and accounts for 60-65% of annual sand flow. In these areas, sand flow from N-NNW directions also occurs and forms 6-10% of the annual total.

The importance of the directional sectors varies seasonally in response to changes in the strength and persistence of different components of the regional wind regime (Fig. 48). The degree of seasonal change is least on the coast, although even here sand flow from easterly directions can account for 30-40% of sand flow in July and August. Inland, the seasonal variability is much greater. Sand flow from S-SW directions dominates in the period September-April and may account for over 90% of sand flow in December-January. In the winter, sand flow from this sector declines and frequently represents only 10% of the total. As the amount of sand flow from the S-SW sector decreases, so the amount from directions between E and NE increases, reaching a maximum of 30-50% in June or July.

The magnitude of potential sand flow, calculated from wind data using the formula of Bagnold (1953), varies considerably over the region (Table 10), in response to variations

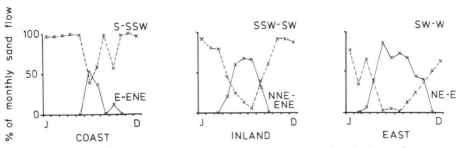

Figure 48. Monthly variation of major potential sand transport directions in the northern parts of the sand sea.

in the strength and persistence of sand moving winds. Total annual potential sand flow is at a maximum in the southern Namib, where it exceeds 1200 tonnes.m.yr^{-1}, reflecting the high energy of the wind regime in this part of the desert. From a maximum in the Luderitz area, annual potential sand flow decreases inland and to the north. In the northern parts of the sand sea, annual potential sand flows are generally much lower, but still exhibit a rapid decrease inland from 440 tonnes.m.yr^{-1} at the coast, to a wide zone where annual potential sand flow is 50-100 tonnes.m.yr^{-1}, and fall to 20-30 tonnes.m.yr^{-1} in the eastern parts of the sand sea. The magnitude of total potential sand flow also varies seasonally. It is at a maximum at coastal stations in November-February and at a minimum in the winter. Inland, maximum total potential sand flow occurs in December-January or locally September, with a subsidiary peak in July.

On the basis of the classification of sand movement regimes devised by Fryberger (1979), the coastal areas of the Namib can be described as high energy, narrow unimodal. Inland in the south, they tend to be wide unimodal or acute bimodal high energy. In the northern part of the sand sea, sand movement regimes are high energy unimodal on the coast and change inland to low to moderate energy obtuse bimodal types, becoming complex in the eastern and possibly also the central parts of the sand sea.

The annual resultant direction of sand flow (Table 10) is almost everywhere towards the NE or NNW, reflecting the dominance of sand flow from the southerly sector. The magnitude of the resultant sand flow is greatest for the high energy, unimodal wind regimes of southern and coastal areas and least for the complex wind regimes of the eastern and northern margins of the sand sea. Seasonal changes in the direction of resultant sand flow are shown in Figure 49. In the summer months, sand flow at all stations is towards the N-NE. The magnitude of both total and resultant sand flow is also at a peak at this time of year. During the winter months, resultant sand flow at most

Table 10. Summary of annual potential sand flow characteristics for DERU and KEP wind recorders.

Station	Sand flow (tonnes m^{-1}.year^{-1}) Total	Resultant	Resultant direction (°)	Undirectional index
AXB	1425	1226	170	0.86
KMK	2312	2024	190	0.88
HLB	671	386	166	0.57
D7	42	22	204	0.53
D8	49	10	223	0.21
NRB	99	46	198	0.46
MD	440	335	196	0.76
CH	54	30	176	0.57
FM	99	56	194	0.57
NN	119	63	213	0.52
MNZ	23	6	278	0.25
RB	278	129	69	0.46
GBB	91	51	65	0.56

For location of stations see Fig. 47

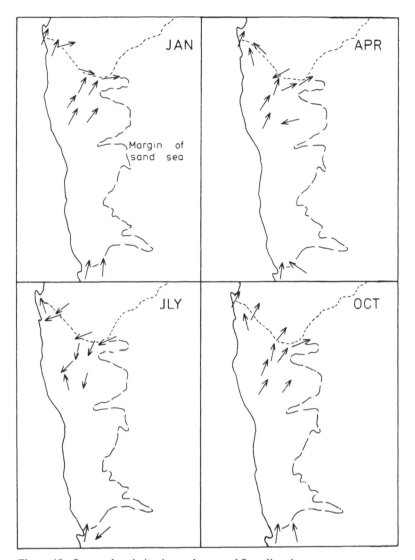

Figure 49. Seasonal variation in resultant sand flow direction.

stations is low and directed towards the SW or W. However, at stations on the northern boundary of the sand sea maximum resultant sand flow occurs in June or July.

The ratio between resultant and total sand flow (the RDP/DP ratio of Fryberger (1979) or the unidirectional index (UDI) of Wilson (1971)) is a valuable index of the effectiveness of sand transport for a station and reliably characterises the sand flow regime. In the Namib Sand Sea, there are three distinct sand flow regimes which can be identified on the basis of their annual UDI (Table 10). The first group comprises coastal stations, with annual UDI's of more than 0.70. The second group consists of stations up to 60 km from the coast, which have annual UDI's between 0.40 and 0.60. The third group consists of stations in the eastern parts of the sand sea, which have indices of 0.20- 0.30. There is a clear relationship (Fig. 50) between the total amount of annual potential sand flow and the UDI, such that there is a division of sand flow regimes into those which have high total and net sand flow and those which have low total and net sand flow.

5.2.2 *Magnitude and frequency of potential sand movements*

Considerations of the magnitude and frequency of events have been widely applied in studies of fluvial and slope processes, but, save for the very preliminary examples in Wolman and Miller (1960) and Warren (1979), have not been applied to studies of aeolian processes. However, use of such concepts can provide new insights into the nature of aeolian sand movements and assist in the interpretation of cyclic crossbedding in aeolian sandstones (Stokes 1964, Hunter and Rubin 1983). Seasonal cycles in the amount and direction of potential sand transport have been described above. It is also evident from autographic records of wind velocity that a quasi-daily cycle of wind velocity also occurs (J.Lancaster et al. 1984). The maximum velocity of S-SW winds occurs in mid to late afternoon, whilst E-NNE and northerly winds often reach their peak in late morning.

Analysis of the magnitude and frequency of sand moving events in the Namib Sand Sea sheds new light upon the distribution in time of sand movements in this region. The following discussion is based upon data for three stations in the northern part of the sand sea with differing wind regime characteristics: high energy unimodal (A), moderate energy bimodal (B) and low energy complex (C). The frequency distribution of winds from all directions and velocity classes and the amount of potential sand flow generated by each wind velocity class are shown in Figure 51.

At station A, which is located on the coast, 80% of total annual potential sand movement is generated by winds with velocities exceeding 8.6 m.sec^{-1}. Winds of this velocity blow for only 6.14% of the time. As reference to Figure 51A will show, the importance of winds with a velocity greater than 11 m.sec^{-1} is considerable. Winds of this

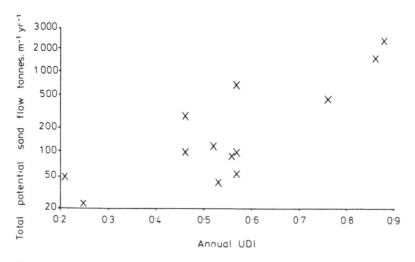

Figure 50. Relationship between the magnitude and directional variability of sand flow.

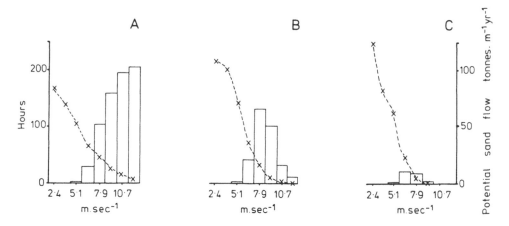

Figure 51. Magnitude and frequency relationships between hours of sand moving winds and potential sand flow; A) high energy unimodal coastal wind regime; B) bimodal moderate energy inland wind regime; C) low energy complex wind regime in the eastern part of the sand sea.

velocity occurred for only 0.87% of the time, yet generated 29% of the annual potential sand flow. At this station, sand flow is dominantly unidirectional, with 79.8% of annual sand flow from the SSW. This is achieved by winds which blow for 33.8% of the time. Of sand flow from this direction, 80% is generated by winds with a velocity between 8.6 and 11 m.sec^{-1}. The dominant effect of winds from one directional sector and moderate velocity class is emphasised by the fact that 34% of annual sand flow at A is generated by SSW winds with velocities between 8.6 and 11 m.sec^{-1}, which blow for only 4.5% of the time. This is equivalent to one hour in every 26.5 hours or approximately one hour a day.

Inland, at station B (Fig. 51B), the situation is somewhat different. Here, the overall wind velocity is generally lower, so that 69% of total annual potential sand flow is generated by winds with a velocity between 7.2 and 9.7 m.sec^{-1}. The annual frequency of such winds is only 4.6%. More striking is the observation that 34% of the total annual potential sand flow is generated by SSW winds with a velocity between 7.2 and 8.3 m.sec^{-1}, which occur for only 2.40% of the time, or one hour every 38.5 hours. The significance of such events in determining overall dune morphology must be very high and they can definately be rated 'formative events'.

At C, on the eastern margin of the sand sea, the overall wind velocity is still less, and no winds with a velocity exceeding 9.7 m.sec^{-1} were recorded. As Figure 51C shows, 89% of total annual potential sand flow was generated by winds with velocities between 5.8 and 8.3 m.sec^{-1}, with an annual frequency of 6%. In a complex wind regime, it is less easy to pick out 'formative' events, but at C, 25% of annual sand flow is generated by SW and WSW winds with a velocity between 5.8 and 8.3 m.sec^{-1}, blowing for 1.66% of the time, or one hour in every 145 hours. Similarly, easterly winds of the same velocity class, with a frequency of only 0.54%, generate 12% of annual potential sand flow.

Analyses of the magnitude and frequency of sand moving winds in the Namib Sand Sea suggest that increased energy of the sand movement pattern is achieved not only by an increase in the amount of time the wind is able to move sand, but by a greater frequency of winds in higher velocity classes, especially those greater than 6.9 m.sec^{-1}. It

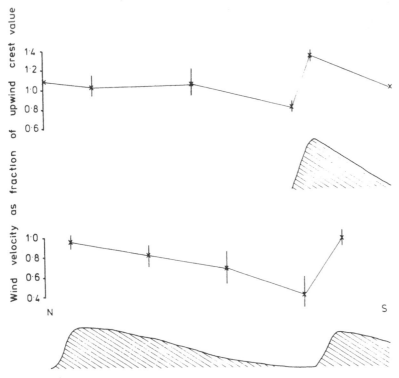

Figure 52. Patterns of wind velocity across isolated and paired crescentic dunes.

is clear that the most effective winds are those with moderate to high energy but relatively infrequent occurrence, especially in low energy wind regimes.

5.2.3 *Variations in winds and sand transport rates across dunes and interdunes*

In addition to regional changes in winds and sand transport rates there are important changes in wind velocity and hence sand transport rates across dunes and interdunes. These have been studied in the Namib Sand Sea on both crescentic and linear dunes (Lancaster 1985b; Livingstone 1986a, b).

5.2.3.1 *Crescentic dunes*
The pattern of wind velocities for winds blowing transverse to the crestline over isolated and paired simple and compound crescentic dunes is shown in Figure 52. In all cases, wind velocities increased steadily up the windward flanks of the dunes. The magnitude of the velocity increases measured by the speed-up ratio (U_2/U_1), where U_2 is the velocity at the dune crest and U_1 is the velocity at the upwind base of the dune, ranged from 1.35 to 1.46 for isolated and 1.79 to 2.25 for paired dunes (Table 11).

The amplification of wind velocity on the windward flanks of dunes is similar to that observed and modelled by meteorologists over natural hills (Jackson and Hunt 1975; Mason and Sykes 1979; Bradley 1980; Pearse et al. 1981). Increases in wind velocity on

Table 11. Mean speed-up ratios between upwind base of stoss slope and crest: crescentic dunes.

Skeleton Coast				
Isolated dunes				
A	1.43			
B	1.35			
C	1.37			
Dune-interdune units				
Interdune-crest		Flanks to crest		
		Lower	Mid	
A	2.24	1.38	1.15	
B	1.99	1.79	1.27	
C	1.79	1.62		
Sandwich Harbour				
Interdune-crest		Mid-flank to crest		
2.36		1.46		

the windward flanks of desert dunes were predicted by Bagnold (1941) and Wilson (1972) and measured but not commented upon by Howard et al. (1978) and Tsoar (1978). They result from the compression of streamlines in the boundary layer as winds pass over the obstacle. The magnitude of the speed-up ratio has been shown by Jackson and Hunt (1975) to follow the relationship

$$\Delta s = 2h/l$$

where h is the height of the hill and l the length measured parallel to the wind at h/2. The speed-up ratio is conceptually equivalent to the amplification factor of Bowen and Lindley (1977) and Jackson (1977).

As Figure 53 shows, there is a clear relationship (r = 0.91 for paired dunes; r = 0.61 for isolated dunes) despite the small sample, between mean speed-up factors and dune shape, as expressed by the ratio 2h/l, for both isolated and paired dunes, although the magnitude of speed-up factors is much greater in the latter case. This accords with the predictions of theory and confirms empirical evidence for wind flow over natural hills (Jackson and Hunt 1975; Bradley 1980). In addition there is a statistically significant (at the 0.05 level) relationship (r = 0.78) between mean speed-up factors and the height of crescentic dunes, as indicated by Figure 53.

Wind velocities drop sharply in the lee of isolated and paired crescentic dunes (Fig. 52). Average velocities at the base of the slip face are 0.40-0.50 of those on adjacent crests for paired dunes and 0.60-0.80 of the equivalent value for isolated dunes. Why this difference occurs is not clear, but it may relate to the existence of significant eddies in the lee of isolated dunes, created by the reattachment of flows around the ends of the dune. In all cases an important component of winds, often very gusty, along the interdune was noted from visual observations and studies of ripple patterns. In the case of isolated dunes, wind velocities recover in their lee to reach magnitudes equivalent to that observed in upwind areas in a distance equivalent to approximately 10 times the dune height.

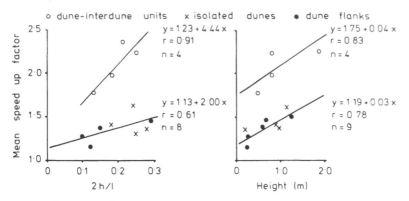

Figure 53. Relationships between mean speed up factors and dune shape and dune height for crescentic dunes.

5.2.3.2 *Linear dunes*

Dominant winds in the area of linear dunes in the Namib Sand Sea are from the SSW and S and cross dunes at an oblique angle. Wind velocity varies by a factor of 1.36-1.80 times across a dune/interdune unit (Fig. 54). There is clear evidence for amplification of wind velocities up western (windward) flanks, followed by deceleration on the upper parts of the east flank. Wind velocities remain fairly constant across lower plinth and interdune areas. In periods of easterly winds, the pattern is similar, but with reversed orientation (Livingstone 1986b). Wind velocities are amplified as they cross east flank dunes and again strongly as they encounter the former avalanche slope. There is a sharp drop in wind velocities in the lee of the crest and an irregular pattern of increasing wind velocity on the western plinth.

Measurements on dunes with heights varying from 3 to 115 m show that, as on crescentic dunes, wind velocities increase steadily up the western slope of the dunes such that the crestal velocity is 1.11 to 2.04 times that on the plinth (Table 12). Unfortunately it was not possible to detect the increased rate of velocity amplification in the upper half of the dune predicted by calculations of amplification factors for triangular and convex hills by Norstrud (1982) and Walmsley and Howard (1985) and linear dunes by Tsoar (1985). However at Narabeb, the positioning of anemometers was sufficiently close to establish that the rate of velocity increase with distance was 1.5-1.65 times greater on the upper third of the western slope of the dune compared to the plinth.

The magnitude of the velocity speed-up varies with both dune shape, as measured by 2h/l, and dune height. The most important influence on the overall magnitude of the velocity speed-up is the wind direction relative to the dune. There is a close correlation between speed-up factors and the direction of the oncoming wind (r = 0.996). Thus the speed up ratio (Δs) is greatest for WSW and W winds and least for SSW and S winds. Mean speed-up ratios averaged 1.65 for WSW winds; 1.51 for SW and 1.33 for SSW, indicating that as the wind blows more parallel to the dune, so its acceleration on the windward flanks becomes progressively less. The degree of velocity acceleration is least for the direction of the wind which is responsible for most sand transport in the area (SSW). This suggests that dune form may be adjusted to such winds, in the manner discussed by Tsoar (1985).

90

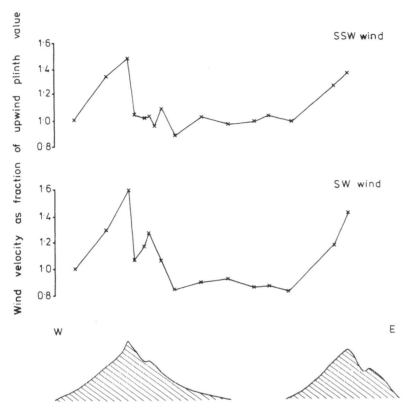

Figure 54. Pattern of wind velocity over linear dunes and interdune areas in periods of SW and SSW winds.

The effects of wind direction on speed up ratios can be eliminated if dune shape (2h/l) is calculated with l measured parallel to the wind direction. As Figure 55 shows, a good correlation exists between mean speed-up factors and 2h/l (r = 0.67, significant at the 0.05 level). A large part of the scatter in this plot can probably be accounted for by the effects of varying dune height, which appears to be more important when winds are blowing at a high angle to the dune. The effects of dune height and shape on speed up factors are closely interrelated. Dune shape effectively changes as wind direction relative to the dune varies. For a given wind direction the shape of most dunes is very similar, although their height varies considerably. This supports the suggestion made above that dune form is adjusted to present least resistance to dominant sand transporting winds.

The amplification of wind velocity on the western flanks of the dune is balanced by a deceleration on the eastern flanks of the dune (Fig. 54). There is an overall decrease in wind velocity such that plinth wind velocity is 0.55-0.60 of that at the dune crest and similar to that of the west plinth of the same dune. Deceleration of S-SW airflow on the east flank of the dune follows an irregular pattern. The largest decrease in velocity occurs in the area of the major slip face such that wind velocity at the base of the main slip face is between 0.50 and 0.70 of that at the crest. This decrease appears to be largely due to flow separation effects.

Table 12. Mean speed-up factors between upwind plinth and crest: linear dunes.

Dune	Wind direction				
	W	WSW	SW	SSW	S
Narabeb					
	1.90	1.60	1.52	1.46	
b	2.04	1.74	1.53	1.32	1.16
c	1.90	1.74	1.34	1.07	
d	1.71		1.47	1.32	
e		1.65	1.42	1.37	
f		1.77	1.65	1.49	
Dune 7					
a		1.89	1.48	1.32	
b		1.63	1.59	1.49	
Dune 8	2.01	1.65	1.47		
Rooibank		1.68	1.65		
Immigration					
a			1.46	1.30	
b				1.37	
Gobabeb South					
a			1.49	1.30	
b				1.11	

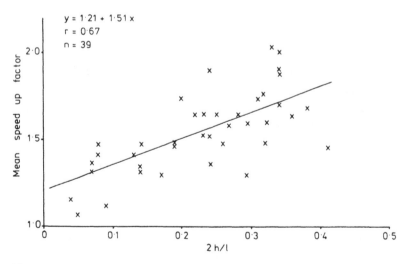

Figure 55. Relationship between mean speed up factors and linear dune shape as represented by the index 2h/l.

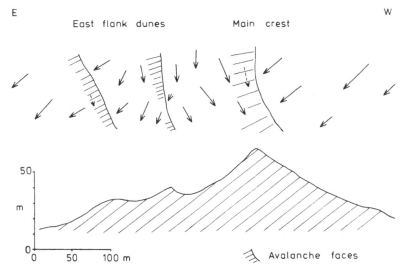

Figure 56. Changes in surface wind direction across a complex linear dune as indicated by wind ripple orientations.

Across the east flank dunes, wind velocities change in a complex manner, with small scale acceleration on windward slopes and deceleration on lee slopes. In general, there is a slight rise in wind velocity away from the base of the main slip face, such that wind velocities on the crests of east flank dunes are 0.75-0.85 of those on the main crest. It appears that the maximum velocity is reached on the crest of the east flank dune nearest the main crest when SW winds are blowing, but on the outermost east flank dune when the main wind direction is from the SSW. In addition the increase in wind velocity away from the base of the slip face is greatest when the overall wind direction is from the SW, indicating that the width of the lee side separation zone varies with wind direction relative to the main crest line, such that the width of this zone decreases as the angle between the wind and the main crestline decreases.

Below the zone of east flank dunes, wind velocity drops sharply, to 0.55-0.61 of that on the main crest, or approximately equal to that on the western plinth. On the eastern plinth wind velocities remain nearly constant or rise slightly by around 10% of their magnitude on the upper part of the plinth. Wind velocities appear to remain more or less constant across the interdune area at a value of 0.50-0.60 that on adjacent crests. Changes in velocity between anemometer stations are less than 10% of the overall plinth and interdune wind velocity.

5.2.3.3 *Changes in wind direction over linear dunes*
In addition to the changes in wind velocities over linear dunes discussed above there are important changes in the direction of surface winds as they cross the dunes, as indicated by flags and surface ripple patterns.

The pattern of wind directions with an overall SW-SSW wind over a complex linear dune is shown in Figure 56. Wind direction remains fairly constant over the western flanks of the dune. However in the zone between the crest and the first of the east flank dunes, winds are diverted such that they blow parallel to the crestline, or locally obliquely towards it and move sand along the dune, in the manner described by Tsoar (1978, 1983). Away from this zone of flow diversion, winds gradually resume their overall direction, but with small scale diversion in the lee of the east flank dunes. On the eastern plinth, wind directions are once again parallel to the overall or mean wind direction.

The pattern of wind directions observed on large complex linear dunes in the Namib Sand Sea is essentially similar to that observed by Tsoar, and indicates that similar processes operate on linear dunes of all sizes. However the zone of lee side air flow diversion extends up to 100 m from the main dune crest, approximately four times the height of the avalanche face, or one tenth the total dune width. This compares with a 5-7 m wide flow diversion zone observed by Tsoar on a dune which averaged 11-13 m in height. In this case, the zone of flow diversion covers most of the lee flank of the dune, whereas in the Namib example, flow diversion affects only the upper parts of the dune. As a result, sand can leave the dunes in periods when wind velocities are high enough to promote sand transport on plinths and thus sand can be transferred from one dune to another (Livingstone 1986b).

5.3 DUNE DYNAMICS

5.3.1 *Rates of dune movement*

Rates of the movement of barchan and crescentic dunes and of the extension of linear dunes have not been measured systematically in the Namib Sand Sea.

5.3.1.1 *Crescentic dunes*

Kaiser (1926) reported that a 33 m high barchan near Bogenfels in the southern Namib coastal zone advanced at an average rate of 48 m.yr[1] during the period May 1916 to March 1919. Gevers (1936) stated that barchans near Luderitz moved at 100-400 paces (80-350 m) a year; and Torquato (1972) gives rates of barchan movement in the Caroca Sand Sea in Angola as high as 100 m.yr[1]. Endrody-Younga (1982) compared the positions of 10 barchans on air photographs of an area SE of Luderitz taken at a 10 year interval. He obtained rates of barchan movement varying from 24.2-60.6 m.yr[1] with a mean rate of movement of 43 m.yr[1].

Compared to rates of barchan movement measured in other arid regions (Table 13) the rates of movement of barchans in the southern Namib are very rapid, considering the size of the dunes investigated. The only similar rates of movement are those documented by Embabi (1982) from the Kharga Oasis, Egypt, but these are for dunes which average 5.5 m high. The high rates of movement observed can be explained in terms of the very high energy of the wind regime in this area. Rogers (1977) calculated potential sand transport rates using the formula of Fryberger (1979) and obtained total drift potential values of 2252 vector units for Bogenfels, between Luderitz and Alexander Bay, with a resultant drift potential of 2065 vector units, higher that any other station in Fryberger's world wide sample of wind regimes in desert regions (Fryberger 1979).

On the northern margins of the sand sea, Barnard (1975) compared the position of 10-13 m high crescentic dunes on air photographs of the Kuiseb Delta region taken in

Table 13. Comparative Rates of Barchan Dune Movement.

	Advance rate (m.yr^{-1})	Dune height (m)
Peru		
Finkel (1959)	15.14	3.67
Hastenrath (1967)	14.2-30.77	3-4
Egypt		
Beadnell (1910)	15	-
Embabi (1983)	48.35	5.52
Algodones Dunes		
Smith (1978)	20 (small) 5 (large)	0.5-6
Salton Sea Barchans		
Long and Sharp (1964)	15-25	5.89
Sinai		
Tsoar (1974)	8.55	2.97
Namib		
Kaiser (1926)	48	33
Endrody Younga (1982)	43	8-10
Barnrd (1957)	8.4	11.33
Ward (1984)	0.8-6.4	10-20

1960 and 1969 and calculated advance rates in this period of 6.8-13.7 m.yr^{1}, with a mean of 9.3 m.yr^{1}. Ward and Von Brunn (1985) measured rates of dune movement for crescentic dunes up to 30 m high on the southern bank of the Kuiseb river between Rooibank and the coast. During the period 1979-1981, these dunes advanced at a rate of 0.80-6.40 m.yr^{1}. Dune movement was generally towards the north or north east with maximum rates of movement in summer months (October- February). During the winter dune movement ceased and some dunes were slightly eroded. Annual rates of dune advance and sand deposition decreased inland from a maximum near the coast, reflecting the sharp decline in wind energy and potential sand transport between coastal and inland areas.

5.3.1.2 *Linear dunes*
Besler (1975) studied the movement of the tip of a 2-4 m high simple linear dune near Gobabeb over a period of 4 years (1969- 1973). She observed a rate of dune extension of 3-4 m per year during this period. At the same time the dune tip also migrated laterally towards the east, giving rise to a net north easterly extension of the dune. Further monitoring of this dune (Besler 1980) established that rates of extension slowed down in the period 1973-1978 to 2 m per year and the tip migrated laterally towards the west.

The rates of extension of linear dunes into the Kuiseb valley between Rooibank and Natab were monitored using aerial photographs and surveyed stakes by Ward and Von Brunn (1985) for the period 1979-1981. Two main zones of dune encroachment were identified. Between Swartbank and Rooibank dune advance rates were minimal (0.08 m.yr^{1} and mostly less than 0.04 m.yr^{1}) with no seasonal trend of movement being evident. Net movement tended to be in a NNW or ENE direction, opposite to directions of potential sand transport computed by Breed et al. (1979), Harmse (1982) and Lancaster (1985a) for stations along the northern margins of the sand sea. However, small shrub coppice dunes on the north bank of the river near Swartbank did exhibit movement towards the south west. Ward and Von Brunn (1985) argued that dune advance rates were minimal in this area because of the bimodal wind regime, with a high proportion of northerly winds. Upstream, between Swartbank and Natab, a second zone of more active dune encroachment was recognised, with rates of dune extension of 0-0.685 m.yr-1. With maximum movement occurring in either summer or winter. The maximum rate of dune encroachment (1.85 m.yr -1) was recorded near Gobabeb, at a site 1 km east of the small linear dune studied by Besler (1975).

Rates of the extension of linear dunes have been infrequently measured. The only comparable measurements are by Tsoar (1978) who recorded a rate of extension of 1.21 m per month or 14.52 m per year for the tip of a 8-10 m high linear dune in Sinai.

5.3.2 *Erosion and deposition patterns on linear dunes*

Rates of dune movement or extension are only one aspect of the dynamics of dunes. The pattern of erosion and deposition patterns on all parts of the dune provides information on how dunes behave under different wind conditions and thus on how they develop. The value of such studies has been demonstrated by Tsoar (1978) and Howard et al. (1978). Erosion-deposition patterns were monitored at biweekly intervals on 3 linear dunes in the northern part of the Namib Sand Sea.

5.3.2.1 *Variations in dune activity*
Variations in the amount of erosion and deposition occur on both temporal and spatial scales. There are changes in the amount of dune activity (the sum of all changes in surface elevation without regard to sign) from week to week as well as changes in the amount of activity across a dune from plinth to crest. The seasonal and spatial patterns of dune activity were essentially similar for each dune studied, despite their varied size (50-92 m high) and different locations.

5.3.2.2 *Seasonal changes*
There is a clear seasonal pattern of dune activity (Fig. 57). Over the period studied, the amount of activity was moderate during October to April, with a decrease from January to April. In late March, there was a slight rise in activity on all dunes, but activity declined sharply thereafter and remained low during April to June. The month from late June to late July was one of very high activity on all the dunes studied with a mean of 20.5% of the annual total dune activity taking place in this period.

Changes in the amount of dune activity follow the seasonal patterns of wind velocity and direction, which give rise to parallel variations in the amount of potential sand transport. Sand transport rates peak in September-January, with around 55% of the total annual sandflow, mostly from S-SW directions, occuring in this period. They decline thereafter to a minimum in April-May which is often a time of weak southerly or northeasterly winds. There is a secondary peak in potential sandflow in June-August which corresponds to a maximum frequency of E-NE winds. The very high rates of dune

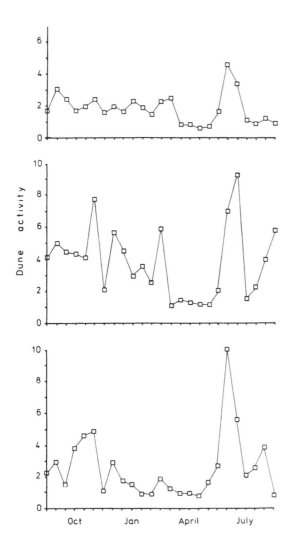

Figure 57. Monthly pattern of dune activity (total of changes in surface elevation without regard to sign) for three dunes studied.

activity recorded in this period correspond to the occurrence of strong to moderate NNE-ENE winds inland and moderate SE-S winds near the coast.

The amount of dune activity in each period is correlated with total potential sand transport (Fig. 58), with correlation coefficients ranging between 0.69 and 0.77. However the amount of dune activity is higher than would be expected in periods when wind directions change from the dominant S-SW direction to either N-NNW or especially E-ENE directions. This is a result of the adjustment of dune form, especially the reversal of slip face orientations, to new wind directions and sand transport patterns. Changes of this nature partly explain the very high rates of dune activity recorded at weeks 12 and 18-21.

5.3.2.3 *Spatial variations in dune activity*
The magnitude of erosion and deposition varies significantly from place to place across

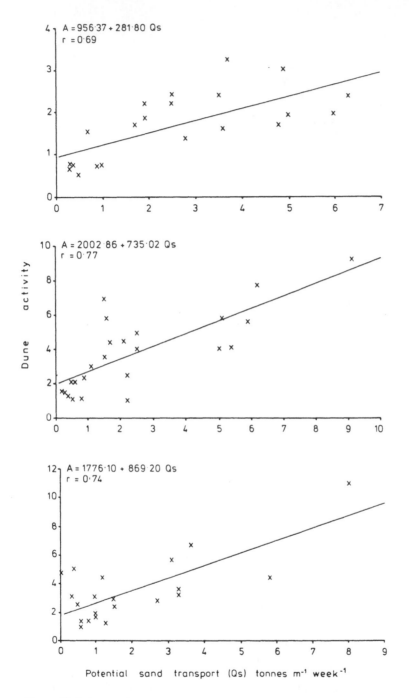

Figure 58. Relationships between annual total of dune activity (total of changes in surface elevation without regard to sign) and total potential sand transport, calculated from winds recorded adjacent to 3 study dunes.

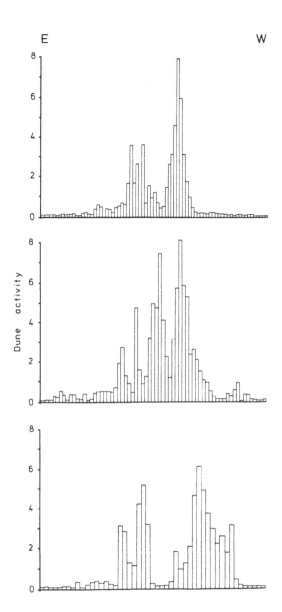

Figure 59. Spatial pattern of annual total of dune activity across the 3 linear dunes investigated.

each linear dune. All dunes display a similar pattern (Fig. 59), which is characterised by low activity on the plinths; a rapid rise in activity up the upper western flanks of the dune to a peak at the dune crest and major slip face; and an irregular decline to plinth values on the east flank of the dune. Between 70 and 91% of the total of erosion and deposition takes place in crestal areas and on east flank barchanoid dunes, with 50-60% of activity in crestal areas and 18-30% on east flank dunes. Dune activity is 19-41 times greater in crestal areas compared with the plinth.

Figure 60 shows the pattern of annual net erosion or deposition across each dune. There is small scale erosion and deposition on both east and west plinths; rapidly

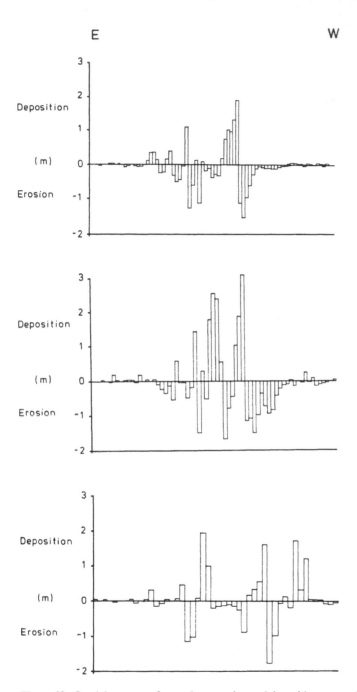

Figure 60. Spatial patterns of annual net erosion and deposition across the investigated linear dunes.

100

increasing rates of erosion on their upper western flanks; net deposition on the major slip face and on the slip faces of east flank barchanoid dunes, equivalent to a net easterly or north easterly migration of these elements of the dune; and an irregular pattern of erosion and deposition on the upper eastern flanks between the superimposed dunes.

Activity on dune plinths is at a low level with an irregular spatial and temporal pattern of small scale erosion and deposition of a few millimeters per week, which probably results from the migration of individual ripples past the measurement points. Where scattered vegetation is present, erosion and deposition rates are somewhat higher, at 5-10 mm per week. This results from localised erosion between plants, or the formation of small shadow dunes in their lee. The shadow dunes so formed are rapidly eroded and reformed when wind directions change, especially from S-SW to E-NE or N-NW.

The crestal regions of the dunes are clearly the most active. In the year over which measurements were conducted the crestlines of all dunes moved towards the east, by an amount varying from 1.9 to 5.2 m. This may be unusual, as a result of the lower than average frequency of NE-E winds in this year. Livingstone (1986a) reports that the crestline of his study dune migrated back and forth over a distance of 14 m, but with little net change over a period of 2-3 years. From October to March, crestlines migrate towards the east or north east, reaching their maximum easterly position in early March (Photo 16A). There is then little change in crest position until the season of NE-E winds, which erode the upper parts of the slip face, and deposit sand as a new slip face on the western side of the crest, which migrates simultaneously towards the west or south west (Photo 16B).

In periods of south to south west winds dune height tends to decrease slightly, with erosion of the crest such that it develops a convex profile (Photo 16A) not unlike that of a transverse dune, in the manner discussed by Tsoar (1985) and Lancaster (1987). With reversal of winds and slip face orientations in the winter, there is a tendency for the crest line to be built up by deposition (Photo 16B).

All dune slip faces recorded net deposition over the year of investigations, indicating a net movement towards the east or ENE of up to 4 m.yr[1] with a mean of 2.13 m.yr [1]. Movement towards the east was strongest in the period October to March with net migration rates over this period averaging 0.38 m.month[-1] (4.61 m.yr[1]). The annual rate of advance is inversely related to the height of the main slip face. The amount of deposition on the slip face increases with the potential sand transport from S-SW directions. Thus, the rate of easterly migration of the main slip faces (Fig. 61) reaches a maximum in the period December-March. During April-June slip faces barely changed position but were severly eroded by ENE-NNE winds in June-August. The amount of erosion was greatest in the upper parts of the slip face.

The upper western flanks of all dunes experienced net erosion over the period of measurement, with the amount of erosion increasing exponentially towards the crest of the dune. The amount of erosion on the upper western flanks varies seasonally, being greatest in the period October-March. There is a positive correlation between the amount of erosion and potential sand transport by SSW and SW winds. Minor deposition occurred during periods of N-NW winds with major deposition by NNE-E winds.

Erosion and deposition patterns on the upper east flanks of the dunes are complex. Often this is the result of infilling or scour of surface irregularities. Most east flank barchanoid dunes recorded a net movement towards the north east, of as much as 5 m.yr[1]

A

B

Photo 16. Reversal of slip face orientation at crest of linear dune; A: summer position; B: winter position. View to North.

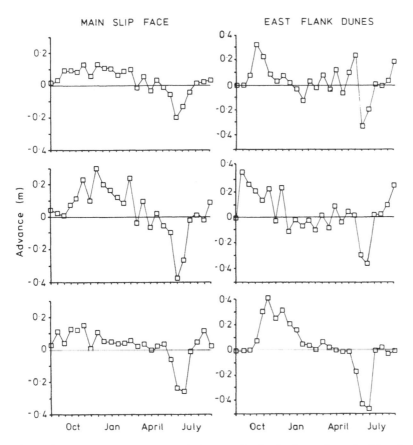

Figure 61. Monthly pattern of rate of advance of main slip face and east flank barchanoid dunes at the three study dunes (negative advance rate idicates erosion).

(mean 2.99 m.yr^{-1}), with the advance rate being inversely related to the height of the dunes. Migration rates were highest, at the equivalent of 2.88-8.21 m.yr^{-1} (mean 5.88 m.yr^{-1}) in the period October-April, with a peak in December- January (Fig. 61). The rate of advance is weakly correlated with potential sand transport from S-SW directions with correlation coefficients ranging from 0.47 to 0.56.

5.3.3 *Origin of the patterns*

The spatial patterns of erosion and deposition observed are related to two main factors: the topography of the dune and changes in wind velocity and surface shear stress over the dune. These are themselves interrelated in a complex and largely unknown manner.

The measurements of wind velocities outlined above show clearly that wind velocities are 1.5 to 2 times higher in crestal areas of dunes compared to the adjoining plinths. Further, for much of the year, the western flanks are zones of accelerating winds, whilst on the eastern flanks winds are decelerating. The erosion and deposition patterns

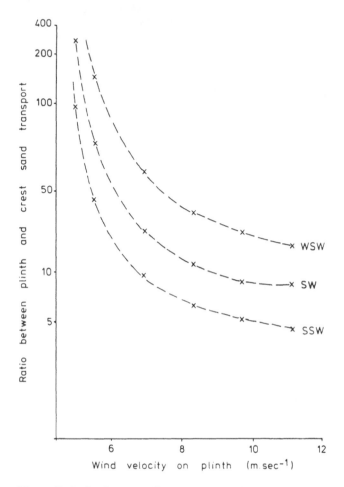

Figure 62. Ration between plinth and crest sand transport rates for representative linear dune (h = 86 m). Note the effect of changing wind velocities and direction on the ratio.

observed are a direct reflection of changes in wind velocity and sand transport rates over the dunes.

As observed by Lancaster (1985b), the major effect of the amplification of wind velocity on the windward flanks of dunes is to increase the amount of potential sand transport towards the dune crest. The magnitude of the increase in potential sand transportation rates varies with both the velocity and direction of the incident wind on the dune plinth and the height of the dune.

In areas of linear dunes it can be observed that, at low wind velocities, there is no sand movement in interdunes or on dune plinths yet sand is moving near the crest of the dunes. Similarly, on lee flanks, sand transport can be observed on the east flank barchanoid dunes and streams of moving sand can be seen to move onto the upper part of the plinth, yet there is no sand movement in the interdune area. In these situations, there is a very large increase in the amount of potential sand transport between plinths and crests. As

104

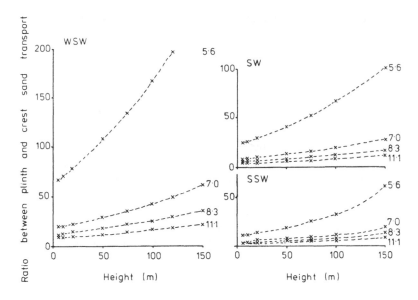

Figure 63. Variation in ration between plinth and crest sand transport rates for linear dunes of different heights and winds from different directions and velocities.

wind velocity on the plinth increases, so the ratio between plinth and crest potential sand transport rates declines exponentially (Fig. 62) from around 200 at just above threshold velocity to 4-15 at a plinth wind velocity of 11 m.sec[-1]. The implication is that as overall wind velocities increase, the overall magnitude of potential sand transport rates rises and they become more uniform over the whole dune. However, there is still a significant increase in sand transport rates from plinth to crest.

As velocity amplification on linear dunes varies with wind direction the ration between plinth and crest sand movement will also vary with wind direction. For a wind which has a velocity of 7.0 m.sec[-1] on the plinth, potential sand transport rates will be 28 times greater at the crest if the wind is from the WSW, but only 8 times greater if it is from the SSW. Similar but less marked differences occur at higher wind velocities. Consequently, potential sand transport rates will be much more uniform over dunes when winds are blowing more nearly parallel to them.

The magnitude of velocity speed-up also increases with dune height, so for a given wind velocity and direction the ratio between plinth and crest potential sand transport rates will vary with dune height. Plinth to crest differences will tend to be at a maximum when winds are blowing at moderate velocities and at a high angle to the dune, as Figure 63 indicates.

For a given dune, varying wind velocities and directions over a period of time give rise to potential sand transport rates in crestal areas which are 8-20 times greater than on adjacent plinths. Differences tend to be greater in periods of light winds and lower in periods of moderate to strong winds. Over the year an eight fold increase in potential sand transport rates between plinth and crest was calculated for Narabeb. This compares with an increase in annual dune activity of 30-40 times between plinth and crest. For periods of uniform S-SW winds the increase in potential sand transport rates between plinth and crest is 9-15 times. This compares with increases in dune activity of 17-23 times.

Bagnold (1941) and Allen (1968) have argued that the amount of erosion and deposition occurring on a dune surface is proportional to the slope angle and the rate of advance of that part of the dune. Thus

$$\frac{\delta q}{dx} = y\, c \tan \alpha \qquad\qquad\qquad\qquad \text{(Bagnold 1941)}$$

where $\delta q/\delta x$ is the rate of sand removal or deposition (dune activity) per unit area at any point. y is the bulk specific weight of dune sand, c is the rate of advance of the dune and α is the slope angle. When $\tan \alpha$ is negative, as on lee slopes, $\delta q/\delta x$ is also negative and thus deposition will occur.

However, this relationship is a purely geometric one, and neither Bagnold (1941) nor Allen (1968) considered the effects of changing wind velocities and sand transport rates over the bedform on the rate of sand removal or deposition, or the rate of advance of the dunes.

5.3.4 A model for erosion and deposition patterns

The effects of changing wind velocities and sand transport rates on the dynamics of dunes can best be explained through the sediment continuity equation (Middleton and Southard 1978; Rubin and Hunter 1982). Thus

$$\frac{\delta h}{dt} = \frac{-\delta q}{\delta x}$$

in which h = surface elevation; q = local volumetric sediment transport in the direction x and t = time.

Using the measured wind velocities over the dune at Narabeb described above, it is possible to evaluate the theoretical pattern and amount of erosion and deposition based upon requirements of sediment continuity. Potential volumetric sand transport rates were derived from Bagnold type equations corrected for the effects of shape using the formula of Howard et al. (1978). The computed patterns of erosion and deposition for SSW and SW winds of different velocities are shown in Figure 64.

Because wind velocities and sand transport rates increase up the windward flanks of dunes, by an amount controlled by the shape and height of the dune, such parts of the dune will be erosional, with the rate of erosion increasing crestwards. The sharp decrease in velocity in the lee of the crest is reflected in massive deposition on the slip face. A varied pattern of localised erosion and deposition occurs across the east flank barchanoid dunes, with the most marked erosion and deposition occurring on the outermost of these superimposed dunes. Both west and east plinths are sites of small scale erosion. Very similar patterns of erosion and deposition were computed by Lai and Wu (1978) for artificial cosine squared shaped dunes (Fig. 65). However, they recorded maximum erosion slightly to windward of the crest of the dune, as did Howard et al. (1978).

Comparison of the computed and actual patterns of erosion and deposition for the dune at Narabeb over a period dominated by SSW and SW winds (Fig. 66) reveals a strong similarity between them. The model predicts the pattern of erosion and deposition on the

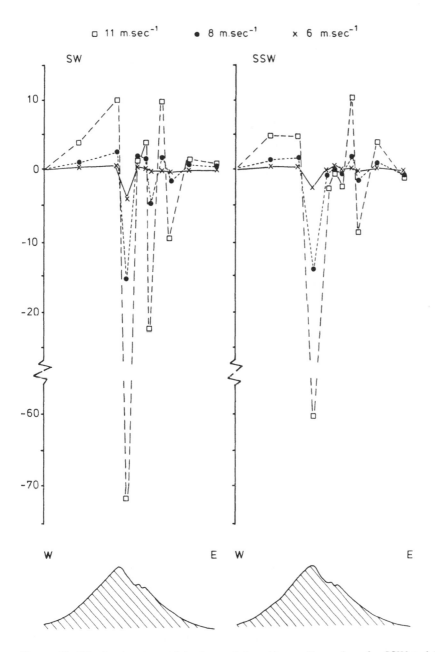

Figure 64. Calculated pattern of erosion and deposition on linear dune for SSW and SW winds of different velocities. Note arbitary units on Y axis.

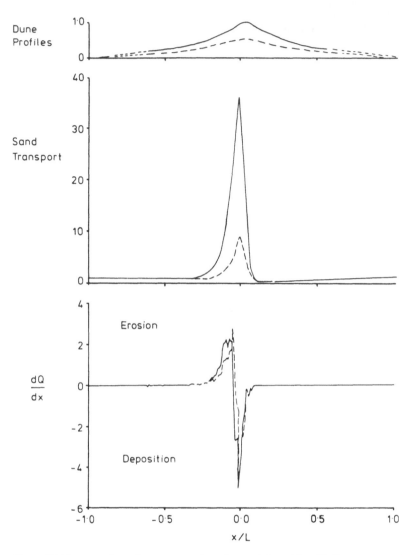

Figure 65. Patterns of sand transport and erosion and deposition ($\delta q / \delta x$) computed for cosine squared shape dune by Lai and Wu (1978).

dune quite well, with erosion on the upper western flanks of the dunes, deposition on the slip face, a low level of erosion and deposition on dune plinths and a complex, mainly depositional zone on the east flank dunes. It would seem therefore that changes in wind velocity over the dune are a major influence on erosion and deposition rates and patterns and hence the dynamics of the dune as a whole. Figure 67 shows that there is a close correspondence between predicted and measured amounts of erosion and deposition, with a correlation coefficient of 0.93. However, the model tends to overestimate the amount of erosion and deposition that occurs on the dune. Howard et al. (1978), using a

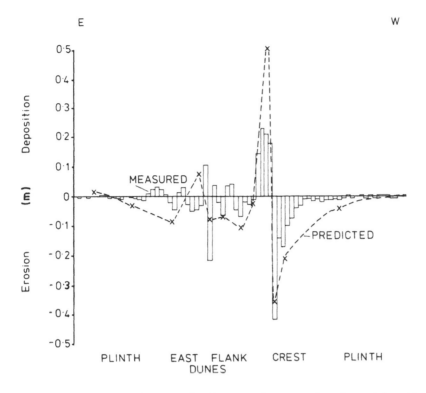

Figure 66. Comparison of measured and computed patterns of erosion and deposition on a complex linear dune during two week period of SSW and SW winds.

similar model, likewise found a close correspondence between measured and simulated patterns of erosion and deposition on a barchan dune.

There is a high degree of interaction between the shape of the dune, the amount of change in wind velocities and sand transport rates and erosion-deposition rates (Tsoar 1985 and Lancaster 1985b, 1987). Given that wind velocities increase up the windward slopes of all dunes, by a varying amount and in a pattern determined by the profile of the dune, it follows that all windward slopes tend to be erosional. Downwind, the wind will have to transport an increasing amount of sand eroded from the dune slope. This in turn requires that wind velocities and surface shear stress should change to increase transport rates proportionately. If they do not, deposition will occur, leading to adjustment of dune form. The shape of the upwind profiles of dunes thus appears to be adjusted to maintain an equilibrium between the rate of erosion and the increase in velocity required to keep an ever increasing amount of sand in transport. On leeward slopes wind velocities decrease rapidly, giving rise to deposition by grainfall and grainflow. On the plinths of linear and star dunes, deposition will occur as a direct reflection of declining sand transport rates.

As discussed by Tsoar (1985), dunes subjected to bidirectional wind regimes tend to have sharp crestlines and a triangular cross section, rather than the convex form of the crest common to most transverse dunes. The triangular profile is the result of the

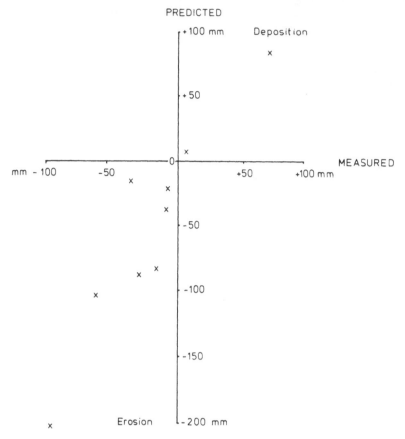

Figure 67. Comparison of measured and predicted amounts of erosion and deposition a on complex linear dune for a two week period of SW and SSW winds.

processes discussed above operating from two directions such that, at a given season, each profile will tend towards adjustment to that wind direction and its spectrum of wind velocities. The high rates of erosion and deposition observed in the crestal areas of linear and star dunes are a reflection of the process of adjustment to changing wind directions.

Dune profiles may be considered to be in a state of transport equilibrium. In practice, such an equilibrium will rarely be achieved, as wind velocities vary diurnally and seasonally. Dune profiles can therefore be considered to be in a state of dynamic or quasi equilibrium (Chorley and Kennedy 1971) with prevailing airflow conditions, in which dune form fluctuates about a series of 'average' states, which may change through time as the dune accumulates. The characteristics of the 'average' state may be determined by the magnitude and frequency of sand moving winds such that, as discussed in section 5.2.2, there are formative events which determine the overall characteristics of dune form in a manner similar in concept to the bankfull discharge in fluvial geomorphology.

110

6 CONTROLS OF DUNE MORPHOLOGY

The morphology of desert dunes is a product of interactions between the sand surface and the wind. The nature of the interactions is modified as the dune grows and projects into the air flow, creating secondary flow patterns. Two main aspects of dune morphology need to be considered: the type of dunes which occur in an area; and their size and spacing.

Six main factors influence the type, alignment, size and spacing of dunes. The nature of dune sands, especially their grain size and sorting characteristics was regarded by Wilson (1972, 1973) as a major influence on the size and spacing of dunes. Sand supply was considered by Mabbutt (1977) and Wasson and Hyde (1983a) to be a determinant of dune type, although satisfactory definitions of this parameter are difficult to achieve (Rubin 1985). The nature of the wind regime, especially its directional variability, has frequently been regarded as the major factor determining dune type. Its effects have been quantified by Fryberger (1979) and Wasson and Hyde (1983a). Wind strength may also be a factor (Ahlbrandt and Fryberger (1980) and vegetation is locally important (Hack 1941; Ash and Wasson 1983). As dunes are the product of an ongoing process of sediment accumulation, the effects of time should also be considered (Mainguet and Callot 1978; Walker and Middleton 1981; Lancaster 1983a). Of these factors, the nature of the wind regime, together with the history of sand accumulation, sediment characteristics and supply are the most important factors determining the morphology of dunes in the Namib Sand Sea.

6.1 FACTORS INFLUENCING DUNE TYPE

6.1.1 *Sediment characteristics*

As discussed in Chapter 4, the evidence for specific associations between dune types and sand grain size and sorting characteristics is confusing. There appears to be no evidence for a genetic relationship between the grain size and sorting character of sands and dune type, except that sand sheets and gently undulating sand surfaces or zibar are often composed of coarse poorly sorted, often bimodal, sands. Similar relationships have been observed in Saharan sand seas (Bagnold 1941; Capot Rey 1947; Warren 1972; Maxwell 1982), Sinai (Tsoar 1978), the Skeleton Coast (Lancaster 1982a) and Algodones dunefields (Nielson and Kocurek 1986; Kocurek and Nielson 1986).

The factors which infuence the formation of low relief aeolian sand accumulations such as sand sheets and zibar are not well known. Kocurek and Nielson (1986) cite five main factors which may be important: high water table, periodic flooding, significant amounts of coarse grained sands, surface cementation and vegetation. The first two factors are probably not relevant in most areas of the Namib Sand Sea.

It is not clear how the presence of significant quantities of coarse sand leads to the formation of sand sheets or zibar. Very coarse grains require a high threshold velocity for movement and are therefore moved infrequently by the wind. This relative stability may permit the growth of vegetation after occasional rains, which further stabilises the surface. In the Namib and Gran Desierto sand seas sand sheet areas are often better vegetated than adjacent dunes. In the Namib Sand Sea this vegetation persists for much longer after rains compared to adjacent areas.

Bagnold (1941) has shown that sand transport by saltation is more effective over surfaces containing coarse grains than over fine sand surfaces. Zibar and sand sheets may form low relief surfaces because they cannot trap sufficient fine sand for the development of dunes. However, Nielson and Kocurek (1986) have shown that small scale flow expansion and deposition does take place in the lee of zibar. Internal structures of sand sheets in the Elizabeth Bay area (Chapter 4) show clearly that sand sheets may accumulate by the deposition of fine sand as shadow dunes and coarse sand by the climbing of granule ripples. In the Great Sand Dunes, Colorado, the presence of vegetation on sand sheets traps fine sand as shadow dunes (Ahlbrandt and Fryberger 1981).

The development of subdued aeolian bedforms such as zibar is probably also closely related to the long saltation paths followed by coarse grains. This is supported by the frequent association of mega ripples with areas of zibar. According to Warren (quoted in Tsoar 1978) the long saltation paths of coarse sands inhibit slip face formation on dunes less than 13 m high.

Wilson (1972,1973) has suggested that different elements of a complex dune pattern may be composed of sands with different grain sizes. Grain size, by controlling the threshold velocity for sand movement, determines the effective wind regime, and therefore may strongly influence dune alignments. However, in most parts of the Namib Sand Sea there is no evidence for this. In areas of complex linear dunes sand from east flank barchanoid dunes and corridor crossing linear dunes is slightly less well sorted and coarser than adjacent linear dune crests, but the differences are not significant in terms of threshold velocities and therefore effective sand transporting wind regimes.

6.1.2 *Wind regimes*

The association of dunes of different morphological types with wind regimes which have different characteristics, especially of directional variability, has been noted by many investigators (e.g. Aufrere 1928; Striem 1954; Tsoar 1974). This has given rise to morphodynamic classifications of dunes as transverse or longitudinal according to the relationship between dune trend and wind direction. However, oblique trends have been widely recognised (Cooke and Warren 1973; Mainguet and Callot 1978; Hunter et al. 1983). In the Namib Sand Sea, as in other sand seas (Cooke and Warren 1973), there are frequently both major and minor alignment trends, whilst in areas of complex dunes

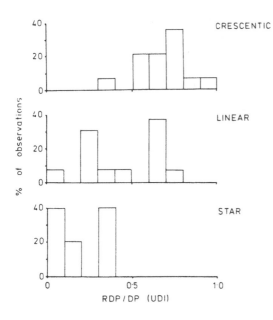

Figure 68. Relationships between dune types and RDP/DP ratios. Data from Fryberger (1979).

multiple dune alignments frequently co-exist. Aufrere (1928), Mabbutt (1968) and Sharp (1966) have argued that each element of crossing dune alignments is the product of a different set of winds in a multidirectional wind regime, each of which builds up a distinct group of transverse or longitudinal dunes. Star dunes may occur at the nodes of these crossing patterns (Wilson 1972). However, Cooke and Warren (1973), following Wilson (1972), have put forward the view that most dune patterns result from a combination of elements which are either longitudinal or transverse to sand transport directions. This often gives rise to an oblique alignment of dunes. In areas of multidirectional winds, selective emphasis of an element of the pattern may result in the emergence of a dominant trend (Wilson 1972). Fryberger (1979) compared dune forms and wind regimes for a variety of sand seas. Wind regimes were characterised by their sand moving potential and classified according to the distribution of potential sand movement directions. The ratio between total (DP) and resultant sand flow (RDP) or RDP/DP was used to give an index of directional variability. This index is equivalent to the Unidirectional Index (UDI) of Wilson (1971) used in Chapter 5. High RDP/DP ratios characterise near unimodal wind regimes whilst low ratios indicate complex wind regimes.

Fryberger (1979) found that the directional variability or complexity of the wind regime increased from environments in which crescentic dunes are found to those where star dunes occur (Fig. 68). Crescentic dunes were found to occur in areas where RDP/DP ratios exceeded 0.50 (mean RDP/DP ratio 0.68) and frequently occurred in unimodal wind regimes, often of high or moderate energy. Linear dunes were observed in wind regimes with a much greater degree of directional variability and commonly occurred in wide unimodal or bimodal wind regimes with mean RDP/DP ratios of 0.45. The trend of linear dunes was observed to be approximately parallel to the resultant direction of sand transport. Star dunes occur in areas of complex wind regimes with RDP/DP ratios less than 0.35 (mean = 0.19).

Table 14. Wind regime environments of different dune types in the Namib Sand Sea.

Dune type and wind recorder (Fig. 47)	UDI	Annual resultant potential sand flow (tonnes m^{-1}yr^{-1})	Wind regime character
Crescentic and barchan			
AXB	0.86	1226	Very high
KMK	0.88	2024	to high energy
MD	0.76	335	unimodal
mean	0.83	1195	
Linear dunes			
Compound			
HLB	0.57	386	High energy obtuse bomodal
Complex			
D7	0.53	22	Low
NRB	0.46	46	to
FM	0.57	56	moderate
NN	0.52	63	energy
RB	0.46	129	obtuse
GBB	0.56	51	bimodal
CH	0.57	30	Low energy wide unimodal
mean	0.52	56.7	
Star dunes			
D8	0.21	10	Low energy
MNZ	0.25	6	complex
mean	0.23	8	

Comparison of the distribution of dunes of different morphological types in the Namib Sand Sea as described in Chapter 3 with the information on sand moving wind regimes contained in Chapter 5 tends to confirm Fryberger's hypothesis that there is an increase in the directional variability of the sand moving wind regime from crescentic to star dunes. However, the overall directional variability of wind regimes in the Namib Sand Sea is less than that in Fryberger's sample.

In the Namib Sand Sea, the spatial distribution of dune types follows the regional variation in sand moving wind regimes. Crescentic dunes occur in areas where winds blow mainly from one direction (e.g. Bagnold 1941; Holm 1960; Norris and Norris 1961). Table 14 and Figure 69 show that in the Namib Sand Sea crescentic dunes of simple and compound varieties are located in areas which experience narrow unimodal wind regimes with a mean UDI index of 0.83 and high to very high potential sand flow. Simple crescentic and barchan dunes occur in areas of high energy wind regimes with compound varieties inland and to the north in areas where wind energy is probably lower. In most areas of the Namib Sand Sea the strike of the crestlines of crescentic dunes is

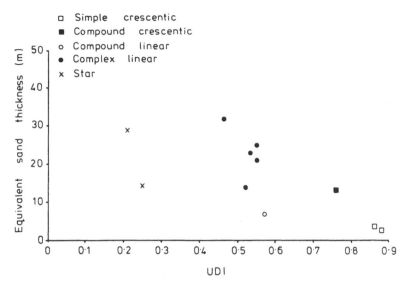

Figure 69. Relationships between dune types, wind regimes and equivalent sand thickness for Namib Sand Sea.

transverse to SSW or SW winds. Alignments swing round inland from transverse to SSW or S winds on the coast and in southern areas of the sand sea to SW inland. This may reflect Ekman spiral effects (Warren 1976) as winds move onto aerodynamically rougher surfaces, or the increasing effects of thermo-topographic circulations on the S-SE geostrophic trade wind circulation around the South Atlantic anticyclone (Rogers 1977).

The origins of linear dunes and their relationship to formative wind directions in the Namib Sand Sea and elsewhere has been the subject of considerable controversy, as reviewed by Lancaster (1982c) and McKee (1982). A widely held view is that linear dunes form parallel to prevailing or dominant wind directions (e.g. Blandford 1876; Aufrere 1928; Madigan 1946; Glennie 1970; Folk 1971). Some investigators (e.g. King 1960; Folk 1971) believe that such dunes are partly erosional, having been formed by material deflated from adjacent interdunes and thus require deep pre-existing sands to form. Such a model was favoured for the linear dunes of the Namib Sand Sea by Barnard (1973). However, Mabbutt and Sullivan (1968) found that the subdune surface continued uninterrupted below the dunes in the Simpson Desert, as it does in the Namib Sand Sea. Evidence from internal structures (McKee and Tibbitts 1964; Tsoar 1982) shows conclusively that linear dunes are basically depositional features.

Bagnold (1953) suggested that heating of air as it passed over desert surfaces would lead to the development of longitudinal roll vortices in which helicoidal flow would sweep sand from interdune areas to dunes. In support of his hypothesis Bagnold pointed to the parallels between dune trends and winds in the hottest months but not the whole year. This model was developed by Hanna (1969) who argued that the expected wavelength of the vortices was 2-6 km, a figure he felt was in general agreement with the observed spacing of linear dunes. Modifications of this model have been widely adopted

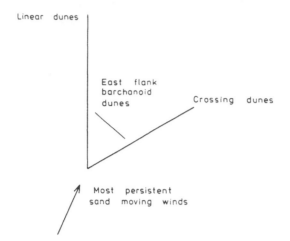

Linear dunes

East flank
barchanoid
dunes

Crossing dunes

Most persistent
sand moving winds

Figure 70. The pattern of alignments in
areas of complex linear dunes.

e.g. Folk (1971), Wilson (1972) and Cooke and Warren (1973). Besler (1977, 1980) adopted the helical roll vortex model to explain the formation of linear dunes in the Namib Sand Sea. She stated that the dunes had been formed by stronger more southerly winds in the period coeval with the last Glacial Maximum and that modern winds were too weak to do more than modify pre-existing forms.

Considerable doubts have been expressed about the validity of the helical roll vortex model of linear dune formation. There are inconsistencies between the spacing of many dunes and the reported dimensions of helical roll vortices (Lancaster 1982c; Livingstone 1986b). Vortices would have to be positioned at exactly the same place at successive wind episodes to allow dunes to grow and extend (Greeley and Iverson 1985), but Angell et al. (1968) have shown that helical roll vortices tend to drift laterally to the mean wind direction, suggesting that they may not be positionally stable over time. However, Tseo (1986) has presented evidence from studies of tethered kites in interdune areas between linear dunes in the Simpson Desert, Australia, which suggest that roll vortices may occur in areas of linear dunes.

There is a substantial body of empirical evidence which indicates that linear dunes form in bidirectional wind regimes and extend approximately parallel to the resultant direction of sand transport. Correlations between dune types and wind regimes (Cooper 1958; Twidale 1972; Clark and Priestley 1980; Fryberger 1979); studies of internal structures (McKee and Tibbitts 1964, Breed and Breed 1979; McKee 1982) and detailed process studies on linear dunes (Tsoar 1978, 1982; Livingstone 1986a, b) support such a view.

In the Namib Sand Sea, most compound and complex linear dunes are found in areas where the annual UDI is in the range 0.46-0.57 (mean = 0.52) and wind regimes are obtuse bimodal, with generally low to moderate potential sand flow (Table 14). Inland, especially in the north and central parts of the sand sea the alignment of the linear dunes is oblique to both major sand flow sectors (S-SW and NE-E) and at an angle of 20-30° to the resultant sand transport direction. These dunes are thus truly oblique in the sense of Hunter et al. (1983). Superimposed barchanoid dunes on the east flanks of these dunes are transverse to resultant sand flows. In some areas there are, in addition, samll simple linear

116

dunes on WSW-ENE alignments which extend from the east flank dunes across interdune corridors (Fig. 70).

The detailed process studies by Tsoar (1978, 1982) show that winds which blow obliquely to the crest at angles of less than 30-50° are diverted to blow parallel to the dune on the lee side and erode and transport sand along the dune. Winds at 50-90° to the crest line will give rise to lee side deposition. Thus any winds from a 180° sector centered on the dune will be diverted in this manner and cause the dune to elongate downwind. This will not necessarily take place in a direction parallel to resultant direction of sand transport, but most probably at an angle of 20-30° to the most persistent sand transporting wind direction.

The process studies of Tsoar (1978, 1982) can be combined with observations of internal structures and wind regimes in areas of linear dunes to produce a general model of linear dune formation (Lancaster 1982c). Linear dunes form in a bimodal wind regime in which the modes are less than 180° apart. They are not necessarily aligned with the resultant direction of sand transport, but frequently form and extend at a small angle (less than 20-40°) to the direction of the most persistent sand moving winds. Thus many dunes of linear form are not longitudinal dunes, but oblique dunes in the view of Hunter et al. (1983) and Rubin and Hunter (1985). Winds blowing at up to 40° to the trend of the dune are diverted to move sand parallel to the axis of the dune on its lee side and are instrumental in extending it. As discussed in Chapter 5, such dune parallel sand movements can be frequently observed on the slip faces and upper eastern flanks of linear dunes in the Namib Sand Sea. Winds blowing at more than 40° to the dune axis tend to give rise to lee side deposition.

The influence of winds from different directions on dune morphology and alignment patterns is therefore related to their persistence and to the angle at which they cross the crest line. This may explain how dunes of different types and on different alignments can co-exist in the same area. In central and northern parts of the Namib Sand Sea the most persistent winds are from the SSW and SW and cross dune trends at optimal angles for dune extension. Further, considerations of the magnitude and frequency of sand movements, discussed in Chapter 5, show that SSW winds of moderate velocity are responsible for 34% of annual potential sand transport in the northern parts of the sand sea. Such winds are instrumental in maintaining and probably determining the overall trend of the linear dunes.

In the central and northern parts of the Namib Sand Sea the S-N linear dunes form the dominant left hand oblique trend. They are at an optimal angle for extension because they lie upwind and therefore receive most sand. The corridor crossing dunes on WSW- ENE alignments form the right hand oblique trend. They lie at much higher angles (50-60°) to dominant sand moving winds and hence experience less extension and more deposition on their lee sides. This would explain why McKee (1982) found a dominance of avalanche face deposits on the northern sides of these dunes. The form of these crossing dunes is also emphasised by N and NW winds in summer, which also extend them. The restriction of this element of the dune pattern to northwestern parts of the sand sea may be related to the regional extent of frequent N and NW winds, as suggested by Besler (1980).

In the southern parts of the Namib Sand Sea most linear dunes are low and of compound form with SE-NW alignments. Although information on wind regimes in

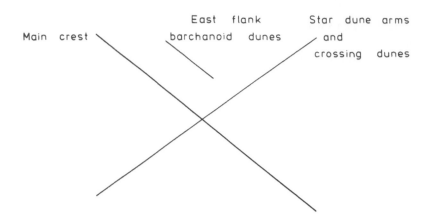

Figure 71. The pattern of dune alignments in areas of star dunes.

these areas is limited, data from the wind recorder at Haalenberg (HLB on Fig. 47), south of the sand sea, suggests that winds here are bidirectional with a dominant SSE-SSW mode (65% of annual sand flow) and a minor ENE mode (15% of annual sand flow). As in the central and northern parts of the sand sea dunes are aligned at a small angle to the most persistent sand moving winds. However, SSE winds which blow nearly parallel to the dune trend are an important component of the wind regime. Such winds will give rise to a small amount of deposition on the dune so that the dunes will be 'sand passing' (Wilson 1972) and thus relatively low.

Star dunes have been regarded as the product of multidirectional or complex wind regimes (Fryberger 1979; Holm 1960; McKee 1966). They may also occur at the nodal points of crossing dune patterns (Wilson 1972). Star dunes in the Namib Sand Sea occur where wind regimes are complex (mean UDI, 0.23) and where both total and net potential sand flow is low. As indicated in Chapter 3, many dunes classified as star types in the Namib Sand Sea do not have a truly stellate form. Commonly, especially in northern parts of the sand sea, there is a preferred orientation to their crestal trend, often paralleling that of nearby linear dunes. The pattern of alignments in areas of star dunes is shown by Figure 71.

Towards the eastern part of the sand sea sand moving wind directions swing around to SW-W. At the same time, the frequency and sand moving potential of E-NE winds also increases such that the sand moving wind regime of the eastern part of the sand sea is opposed bimodal or locally complex. This constitutes a limiting case of the model of linear dune formation outlined above, when winds cross dune trends at around 90°. In these situations, the tendency for dune extension is minimal. Sand remains on the dune and is eroded and redeposited on opposite sides of the crestline, resulting in upward growth by deposition, rather than extension. Many of the star form dunes on the eastern edge of the sand sea are thus effectively reversing dunes, especially where topographic funnelling of winds by valleys leading from the escarpment occurs, as for example at Sossus Vlei.

Along the north edge of the sand sea close to the Kuiseb valley and in the southern part

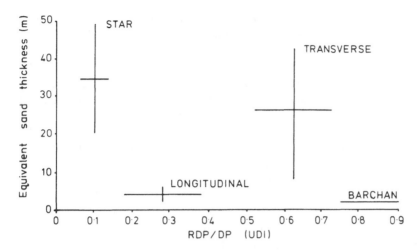

Figure 72. Relationships between dune types, wind regimes and equivalent sand thickness. After Wasson and Hyde (1983).

of the sand sea, many star dunes have a more truly stellate or pyramidal form. This probably results from the occurrence of a distinctly tri-modal wind regime in these areas. Along the Kuiseb valley, in addition to persistent moderate strength SSW-SW winds strong upvalley N-NW winds are recorded on summer mornings (Tyson and Seely 1980) and strong E- NE winds frequently occur during the winter (J.Lancaster et al. 1984). In the southern part of the sand sea north westerly winds in winter are associated with the passage of mid latitude depressions to the south of the region and give rise to a complex wind regime at Aus (Fryberger 1979).

Subsidiary trends in areas of star dunes (Fig. 71) indicate that crests of barchanoid dunes on the flanks of star dunes are transverse to SW-WSW winds; whilst are arms of star dunes extend at a small angle to SW winds and would therefore be extended by all winds from this sector.

6.1.3 Sand supply

The availability of sand for dune building has long been considered a factor influencing dune morphology (Hack 1941; Wilson 1972). Recently, Wasson and Hyde (1983a) have attempted to quantify the relationships between sand supply, wind regimes and dune morphology. Sand supply was defined in terms of equivalent sand thickness (EST) or the spread-out thickness of dune sand in a given area. Using a sample of dunes of different types from sand seas in all major desert areas, Wasson and Hyde (1983a) established that the mean EST for transverse and star dunes and also for linear and barchan dunes was statistically identical and concluded that although EST was a significant variable determining dune type, it was not the only one.

However, by plotting EST against Fryberger's RDP/DP ratio, a clear discrimination of dune types was achieved (Fig. 72), leading to the conclusion that barchans occur where there is very little sand and almost unidirectional winds: transverse dunes where sand is abundant and winds variable; linear dunes occur where sand supply is small, but winds more variable still; and star dunes are found in complex wind regimes with abundant sand supply.

By using data from surveyed cross sections of dunes, it is possible to determine the mean EST for dunes of different types and varieties in the Namib Sand Sea (Fig. 69). Comparisons between Figures 69 and 72 indicate that all dune types in the Namib Sand Sea occur in wind regimes which are less variable in direction than Wasson and Hyde's sample. This is especially true for linear and cresentic dunes. In addition, star dunes in the Namib Sand Sea occur where EST is lower than the Wasson and Hyde model, and linear dunes, especially of complex varieties, where EST is higher than predicted. In the Namib Sand Sea there is a increase in EST as wind regimes become more variable (Fig. 69).

Rubin (1984) has raised a number of important criticisms of Wasson and Hyde's use of EST as a measure of sand supply. In particular, EST is actually not a measure of sand supply, but of the volume of sand contained in the dunes and may be a reflection of dune type with the dune type being influenced by other factors, especially the wind regime. In the Namib Sand Sea it is possible to clearly discriminate between dune types on the basis of their relationships with wind regimes alone. The EST data merely suggests that there is more sand in complex linear and star dunes than in compound linear and all types of crescentic dunes.

6.2 NATURE OF THE RELATIONSHIPS BETWEEN DUNE HEIGHT AND SPACING

The close relationships between dune height, width and spacing have been documented for dunes of different types in the Namib Sand Sea in Chapter 3. Similar relationships exist in sand seas elsewhere (Wilson 1972; Breed and Grow 1979; Lancaster 1982a; Wasson and Hyde 1983b; Lancaster et al. 1987). The available data is summarised in Figure 73.

Although most of the relationships between dune height, width and spacing have been published as arithmetic plots, their general form can also be expressed by a power function of the form

$$D_H = c\, D_S^{\,n} \qquad (1)$$

where D_H is dune height, D_S is dune spacing and the exponent n is a measure of the rate of change of the dependent variable relative to the rate of change of the independent variable.

Variations in dune height and spacing over the sand sea can be examined using the concept of allometry or the analysis of the relative rates of change of two parts of a system. The concepts of allometry are widely used in the biological sciences (e.g. Gould 1966) and their potential for understanding depositional landforms has been discussed by Bull (1975). Examination of dune height/spacing relationships in these ways is an example of dynamic allometry in which changes in time are examined using spatial data (Bull 1975).

It is clear from Figure 73 that the slope of the regression lines and therefore the values of the exponent of the power function vary between sand seas, as well as from one dune

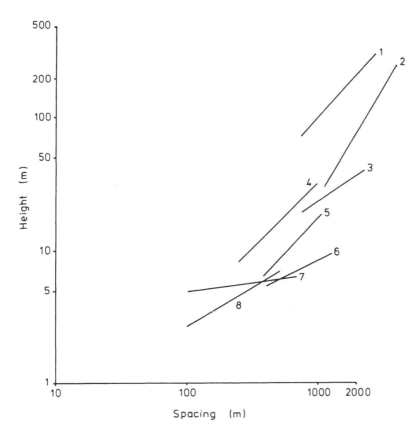

Figure 73. Relationships between dune height and spacing. Data for Australian sand seas from Wasson and Hyde (1983). 1) Namib Sand Sea star dunes; 2) Namib Sand Sea complex linear dunes; 3) Namib Sand Sea compound linear dunes; 4) Namib Sand Sea crescentic dunes; 5) Simpson Desert simple linear dunes; 6) Great Sandy Desert simple linear dunes; 7) Skeleton Coast dunefield crescentic dunes; 8) Gran Desierto crecentic dunes.

type to another in the Namib Sand Sea. In addition, the value of the constant term varies by over 300 times. Thus the crest to crest spacing of a 5 m high simple crescentic dune would be 100 m in the Skeleton Coast dunefield but 150 m in the Namib Sand Sea and 300 m in the Gran Desierto sand sea. Similarly, 10 m high simple linear dunes are 490 m apart in the Simpson-Strzelecki sand sea, but 1300 m apart in the Great Sandy Desert. In the Namib Sand Sea, compound linear dunes 2000 m apart will be 37 m high, yet complex linear dunes of the same spacing will have a height of 70 m and star dunes with the same spacing will be 180 m high. Variations in the value of the constant appear to reflect differences in the amount of sand which has been incorporated in the dunes. In areas of crescentic dunes, variations in the constant may also be an indication of changes in dune steepness, which in turn is controlled by grain size and the distribution of wind velocities (section 6.3.1).

The exponents of the power function in Equation 1 vary from 0.52 to 1.72 (Table 15).

Table 15. Exponents of power functions of dune height/spacing relationships in Figure 79.

Namib Sand Sea	
Crescentic dunes	0.97
Compound linear dunes	0.54
Complex linear dunes	1.72
Star dunes	1.20
Skeleton Coast Dunefield crescentic dunes	1.40
Gran Desierto crescentic dunes	0.58
Simpson-Strzelecki Sand Sea linear dunes	1.06
Great Sandy Desert linear dunes	0.52

Some areas or types of dunes follow positive allometric relationships, whilst others display negative allometry. Strong positive allometry is shown by complex linear dunes and star dunes in the Namib Sand Sea and crescentic dunes in the Skeleton Coast dunefield. Linear dunes in the Simpson-Strzelecki Desert and crescentic dunes in the Namib Sand Sea are close to isometry whilst compound linear dunes in the Namib Sand Sea, linear dunes in the Great Sandy Desert and crescentic dunes in the Gran Desierto display moderate negative allometry.

There are three possible allometric models to explain dune size/spacing relationships: (i) Isometry: dune height increases at the same rate relative to dune spacing such that the exponent of the power function equals unity. Thus a given amount of sand can be formed into a few widely spaced dunes or many small closely spaced dunes. Such a model was advocated for Simpson Desert linear dunes by Twidale (1972). (ii) Positive allometry: dune height increases more rapidly than dune spacing, indicating a tendency for vertical growth of the dunes. The exponent of the power function is greater than unity. (iii) Negative allometry: dune height increases less rapidly than dune spacing. Dunes grow towards an 'equilibrium' height and maintain that height thereafter. The exponent of the power function is less than unity.

In the Namib Sand Sea, concepts of allometry suggest that crescentic dunes and compound linear dunes grow towards a maximum height at an asymptotic rate, whilst complex linear dunes and star dunes show a strong tendency for vertical growth without an equivalent change in dune spacing. Positive allometry in dune height-spacing relationships of linear and star dunes is a reflection of wind regimes which favour deposition on dunes and hence their vertical growth. It also suggests that dune spacing may be determined before dune height and by some other mechanism. Negative allometry in dune height/spacing relationships represents a situation in which dunes are growing towards a height which is in equilibrium with current sand supply and wind regime characteristics. Further change takes place by dune extension or migration. In the Namib Sand Sea dune types exhibiting negative allometry occur in the southern and western parts of the sand sea where total and resultant potential sand transport rates are higher compared with those in areas of complex linear and star dunes. Thus negative allometry may characterise dunes in zones of active throughgoing sand transport and rapid dune extension or migration.

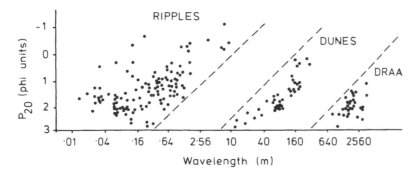

Figure 74. Relationship between bedform wavelength (spacing) and P_{20}. After Wilson (1972).

6.3 FACTORS INFLUENCING DUNE SIZE AND SPACING

The pattern of dune size (height, width) and spacing in the Namib Sand Sea as documented in Chapter 3, displays clear evidence of systematic organisation, with the largest and most widely spaced dunes in central and northern areas of the sand sea and smaller and more closely spaced dunes towards the margins and in the southern part of the sand sea. The close correlations which exist between dune height, width and spacing for each dune type imply a high degree of adjustment to controlling variables (Chorley and Kennedy 1971).

The nature of the variables which control dune size and spacing are poorly understood. In subaqueous environments, bedform spacing increases with flow depth and grain size (Allen 1968). If flow depth and grain size are held constant, then bedforms increase in size as flow strength increases (Yalin 1972; Kennedy 1969). A downcurrent decrease in transport rate will consequently lead to a decrease in bedform size or number of bedforms (Rubin and Hunter 1982).

Ultimately, the limit on dune size is sand availabilty. Dunes cannot grow unless sufficient sand is available for dune building.

6.3.1 *Grain size*

Wilson (1972) plotted the grain size of the coarse twentieth percentile (P_{20}) of dune crest sands against the spacing of aeolian bedforms and found a clear separation into a hierarchy of ripples, dunes and draas (Fig. 74). No transitional forms were observed, suggesting that, in particular, dunes did not evolve into draas. The spacing of dunes and draas was thought to be related to the size of secondary flow elements which were in turn related to wind velocities. The grain size of P_{20}, according to Wilson (1972), closely approximates the mean size of surface sand in mixed deposits. Wilson argued that changes in dune spacing were caused primarily by changes in the grain size of source sediment. For a given grain size, there is a minimum shear velocity which will move this sand and hence a minimum possible dune spacing.

Using the data base on grain size and dune spacing for the Namib Sand Sea, it is possible to test Wilson's hypothesis, as has been done for Australian dunefields by

123

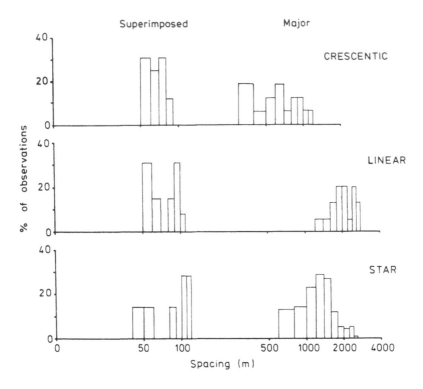

Figure 75. Hierarchical arrangement of dune spacings in the Namib Sand Sea.

Wasson and Hyde (1983b). Excluding ripples, for which there is no data, there is a clear hierarchy of bedform spacings in the Namib Sand Sea as shown by Figure 75. The two orders of bedform spacings in areas of complex and compound linear and star dunes correspond directly to that of the main dunes and the subsidiary or superimposed bedforms on their flanks. This is a true bedform hierarchy in the sense of Allen (1968). In areas of crescentic dunes, the smaller spacings similarly correspond to that of superimposed bedforms, but the more widely spaced dunes include both simple and compound forms. However, consideration of major dunes only shows that there is an overlap between the spacings of each type and no clear division into dunes and draas. In fact, spacings of linear dunes in the Namib Sand Sea fit into the gap between dune and draa spacings on Wilson's plot.

A plot of mean dune spacing against P_{20} for the Namib Sand Sea (Fig. 76) similarly does not clearly discriminate between orders of bedform spacing. Similar conclusions were reached by Wasson and Hyde (1983b) using data for Australian dunefields. They also found that there was no relationship between dune spacing and P_{20} for linear dunes. In the Namib Sand Sea, there is similarly no relationship between spacing and P_{20} for linear and star dunes (Fig. 76). However, there is a statistically significant relationship between the spacing of crescentic dunes and P_{20} ($r = 0.54$).

In agreement with Wasson and Hyde (1983b) it would appear that Wilson's hypothesis of a grain size control of dune spacing is not universally applicable. There appears to be

124

no relationship between P_{20} and the spacing of linear and star dunes either in the Namib Sand Sea or Australian dunefields. However, data from the Namib Sand Sea, the Skeleton Coast dunefield (Lancaster 1982a) and Gran Desierto Sand Sea (Lancaster et al. 1987) indicates that there is probably some grain size control of the spacing of crescentic dunes (Fig. 77). This suggests that the formative mechanisms for linear and crescentic dunes may be different and their spacing is influenced by different controlling variables.

One possible mechanism to explain the spacing of crescentic dunes in terms of grain size and wind velocities was put forward by Lancaster (1985b), who argued that the dimensions and shape of simple dunes in unidirectional winds can be evaluated using the sediment continuity equation (Middleton and Southard 1978; Fredsoe 1982; Rubin and Hunter 1982). Thus

$$\frac{\delta h}{\delta t} = \frac{-\delta q}{\delta x} \qquad (2)$$

where h is local bed elevation, q is local volumetric sediment transport in the direction x and t is time. Each bedform is migrating at a rate Q_B given by

$$Q_B = \frac{q \ (crest)}{H} \qquad (3)$$

where H is the height of the dune. Combining the two equations (Fredsoe 1982) gives

$$\frac{h}{H} = \frac{q \ (local)}{q \ (crest)} \qquad (4)$$

It is possible to calculate q (local) for any point on the dune using the data on velocity amplification discussed in Chapter 5, and from this to evaluate the shape of transverse bedforms which will be in equilibrium with a given wind velocity. Examples of these are given in Figure 78. They suggest that at low wind velocities dunes tend to develop rather steeper profiles compared to those developed at high velocities. Transverse dune ridges become broader and more rounded as wind velocities increase, in a similar manner to subaqueous bedforms (Yalin 1972; Fredsoe 1982). This suggests that the shape and also the spacing of such dunes for a given height is governed ultimately by the overall wind velocity in an area. Thus dunes are shorter, steeper and more closely spaced in areas of low average wind velocity and longer and spaced further apart in areas of high winds.

Sediment size, through its influence on threshold velocities and transport rates, is also important. Lower wind velocities will be required to move sand on fine grained dunes, which will develop a similar morphology to those created by low wind velocities with short, steep profiles and therefore a close spacing. Coarse sands may therefore result in broad dunes with a large crest to crest spacing for their height. This model may be a partial explanation of the observed relationships between crescentic dune spacing and P_{20}.

6.3.2 Aerodynamic models of dune size and spacing

Suggestions that the spacing of dunes is related to the scale of secondary flow circulations

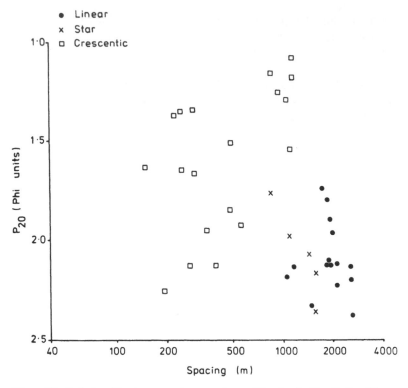

Figure 76. Relationships between dune spacing and P_{20} for the Namib Sand Sea.

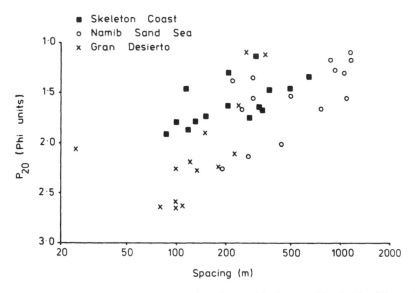

Figure 77. Relationship between spacing of crescentic dunes and P_{20} in Namibian and Gran Desierto sand seas.

126

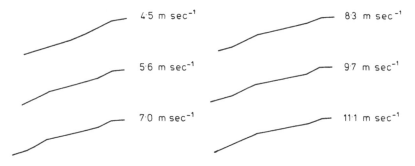

Figure 78. Simulated transverse dune profiles at different wind velocities.

or to the dimensions of the zone of flow disturbance downwind of dunes have been put forward by a number of investigators (e.g. Hanna 1969; Folk 1971; Wilson 1972; Twidale 1972; Tsoar 1978; Lancaster 1983b). In particular, Hanna (1969) and Folk (1971) related the spacing of linear dunes to the scale of horizontal roll vortices and hence the depth of the planetary boundary layer. Yalin (1972) and Cooke and Warren (1973) suggest that the scale of natural atmospheric turbulence is a major influence on dune size and spacing. Howard et al. (1978) have argued that possible controls of dune size include upwind roughness, with dune size increasing with the size of fixed roughness elements such as rocks or vegetation.

Tsoar (1978) and Lancaster (1983b) suggested that the spacing of linear dunes was related to the distance over which airflow was disturbed in their lee. This model receives some support from studies of subaqueous bedforms (Engel 1981; Fredsoe 1982) which suggest that the flow separation zone extends down current for 5-10 times dune height. In the Sahara, Knott (1979) noted that turbulent intensity remained high for distances of 300-400 m downwind of small barchan dunes. As noted in chapter 5, wind velocities in the lee of isolated crescentic dune ridges recover to their upwind value by a distance equivalent to 10 times the dune height. Studies of winds downstream of natural and artificial obstacles suggest that wind velocities are reduced for a distance up to 12-15 times the obstacle height (Chepil and Woodruff 1963; Oke 1978). Tsoar (1978) suggested that the spacing of linear dunes was approximately 12-15 times their height. However, as pointed out by Lancaster (1983b), winds cross these dunes at an oblique angle and their spacing is much greater than this along the line of the wind. In the Namib Sand Sea, the distance between linear dunes along the line of the most persistent sand moving winds (SSW) averages 23 times dune height.

Measurements of the spatial variation in wind velocity across and between linear dunes discussed in Chapter 5 show that the zone of disturbance of winds does not extend across interdunes, where wind velocities remain essentially constant. This implies that separation zone and wake models may not be appropriate for explaining the spacing of large linear dunes.

6.3.3 *Sand transport rates*

By analogy with subaqueous bedforms, it might be expected that dune size is related to

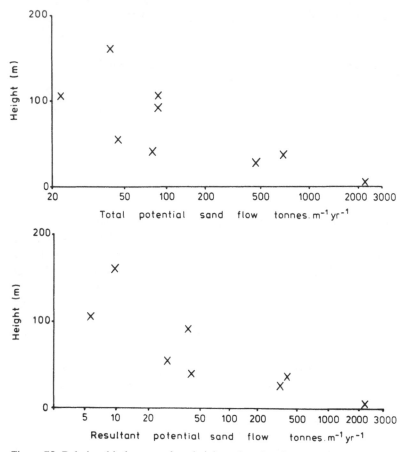

Figure 79. Relationship between dune height and total and resultant sandflow in the Namib Sand Sea.

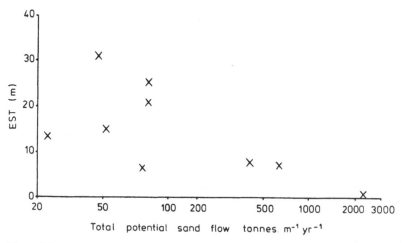

Figure 80. Relationship between equivalent sand thickness and total potential sandflow in the Namib Sand Sea.

128

overall sand transport rates such that larger dunes would be located in areas of high wind energy, as represented by high total potential sand transport rates; and small dunes in areas of low wind energy. However plots of mean dune height against both total and resultant potential sand transport for stations in the Namib Sand Sea (Fig. 79) show that exactly the opposite situation occurs: large dunes are found in areas where potential sand transport rates are low and small dunes in areas of high sand transport rates. Similarly, equivalent sand thickness, which can be regarded as an index of the total amount of sand in the bedforms of a given area, decreases in a linear fashion with increasing total potential sand transport rates (Fig. 80).

Although dune form in the Namib Sand Sea appears to be in general equilibrium with the directional characteristics of the wind regime, the pattern of dune size and spacing is not in accordance with the predictions of models of bedform development. Two explanations for this may be put forward.

(i)   Small bedforms may occur in areas of high potential sand transport because the actual sand flow is less than potential sandflow (metasaturated sand flow in the terminology of Wilson (1971)). Thus, small bedforms in some areas may be a result of low sediment concentrations and a poor sand supply.

(ii)   Simple and compound and complex dunes are genetically distinct bedforms.

Data from surveyed cross sections of linear and star dunes in the Namib Sand Sea shows that their cross sectional area increases exponentially with dune height (Fig. 81). Large dunes of these types contain proportionately much more sand than do smaller examples. When the effects of variable dune spacing are taken into account, there is a strong linear relationship between equivalent sand thickness in an area and dune height (Fig. 82). It is clear that large complex dunes, even though widely spaced, represent a net accumulation of sand.

Despite the arguments of McKee (quoted in Kocurek 1981b) that complex and compound dunes are merely larger versions of simple dune forms, there is a large body of evidence from the rock record (Brookfield 1977; Kocurek 1981a, b; Hunter and Rubin 1983; Rubin and Hunter 1985; Mader and Yardley 1985) which indicates that large aeolian bedforms (draa or complex or compound dunes) are built up by the migration of smaller dunes across them, creating sets of cross strata separated by second order bounding surfaces. Following the hierarchy of bounding surfaces defined by Brookfield (1977), the sediments of simple dunes will contain only 3rd order surfaces, whilst complex and compound dunes will contain both 2nd and 3rd order surfaces.

In the Namib Sand Sea, as elsewhere (Wilson 1972), a hierarchical system of aeolian bedforms exists (Fig. 75). Following Jackson (1975) three orders of bedforms can be recognised: (i) micro forms : ballistic ripples; (ii) meso forms : individual simple dunes or superimposed dunes on compound and complex dunes; (iii) macro forms : compound and complex dunes (draa of Wilson 1972). Each of these bedform orders has a characteristic time period, termed the relaxation or reconstitution time, over which it will adjust to changed conditions (Allen 1974). In the case of aeolian bedforms, relaxation time increases by up to several orders of magnitude from micro to meso and macro forms. Because change in bedforms involves movements of sediment, an increasing spatial scale is involved at each level of the hierarchy. Thus ripples, the smallest bedform in the hierarchy, have a morphology which is controlled by the nature of individual dynamic events (periods of sand moving winds) and a short life span (hours or days).

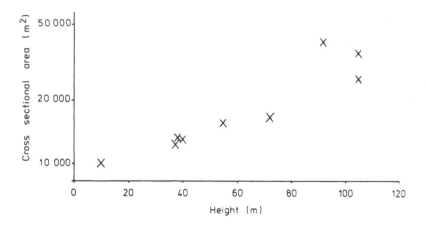

Figure 81. Relationship between cross sectional area of dune and dune height for surveyed dunes.

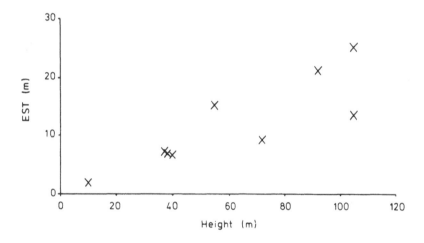

Figure 82. Relationship between equivalent sand thickness and dune height for surveyed dunes.

The morphology of simple dunes is governed principally by seasonal scale patterns of wind direction and velocity and by spatial changes in winds over macro forms. Their life span is in the order of 10^0 to 10^2 years. Analogous subaqueous bedforms are mega ripples and sand waves in tidal environments. The largest aeolian bedforms, complex and compound dunes, respond only to changes in the overall geomorphological regime (sensu Jackson 1976). They are relatively insensitive to changes in local flow conditions and may persist for 10^3 to 10^4 years. Macro scale dune forms or draa are comparable in scale to point bars and sand banks in shallow seas.

This difference of scale suggests that the factors which control the size and spacing of simple dunes should be considered separately to those which influence the size and spacing of complex and compound dunes or draa. Whilst the height and spacing of small, simple dunes is probably related to contemporary sand supply and sand flow regimes, the

130

size and spacing of large complex and probably compound bedforms is a function of the long term accumulation of sand in certain areas of the sand sea determined by regional scale patterns of winds and sand transport. Their distribution in a sand sea therefore reflects the spatial pattern of deposition and their size is a result of long continued growth in conditions of abundant sand supply.

6.4 THE DEVELOPMENT OF AEOLIAN BEDFORMS

Bedforms develop primarily because sediment transport across flat surfaces in conditions of fluctuating flow is dynamically unstable. The existence of bedforms is analogous to the formation of meanders in rivers (Yalin 1972) and represents the effects of conservation of energy expenditure in the transporting medium. Variations in wind velocity and sand transport rates occur, at a scale significant for dune and draa formation, on a diurnal and seasonal basis. In addition there are significant changes in sand transport rates over dunes, as discussed in Chapter 5. Following Kennedy (1969) bedform equilibrium is reached when the local transport rate varies linearly with local bed elevation. Because of the fluctuation of wind velocity with time and the lag between changes in transport rates and dune form (Allen 1980) this equilibrium will be at best partial.

Given a perturbation in the sand transport rate, lag effects between changes in flow velocity and transport rates will give rise to deposition (Kennedy 1969). In many desert areas, diurnal variation of wind velocity may be an important factor in initiating bedforms by deposition. Bedforms, once initiated, will tend to enhance deposition and erosion patterns through local flow convergence and divergence. This may lead to the selection of certain, probably larger, bedform wavelengths for rapid growth and the suppression of other shorter wavelengths (Huthnance 1982), giving rise to a characteristic spacing of bedforms which tends to be self maintaining. In part the hierarchical arrangement of bedform size and spacing may be seen as a harmonic pattern of 'selected' wavelengths.

Seasonally oscillating flows, i.e. bidirectional or multidirectional wind regimes, will similarly tend to select and enhance by flow convergence certain wavelengths and trends of a pattern and subdue others, leading to the formation of linear and star dunes. Thus dunes of these types may lie at the convergent points of seasonal patterns of sand transport. Because sand transport to dunes is relatively more effective than transport from dunes, as a result of lee side flow separation and diversion and deceleration, sand once transported to the dune will tend to stay on it and hence enhance its growth.

The size and spacing of dunes may be seen as the product of variations in sand transport rates at various scales: the sand sea, the dune and the dune-interdune unit. In turn these are the result of equivalent spatial changes in wind velocity and directional variability.

7 THE ACCUMULATION OF THE SAND SEA

Sand seas are the largest aeolian sand bodies and contain considerable volumes of aeolian sand, together with local but significant amounts of interdune and extra-dune fluvial lacustrine and marine sediments (Lupe and Ahlbrandt 1979). Their accumulation probably requires long periods of geological time. Estimates of the time required for modern sand seas in the Sahara to accumulate range from 1.35 to 2 Ma, but up to 50 Ma may be required to reach full equilibrium with sand flow conditions (Wilson 1971). On this time scale, tectonic effects and basin subsidence will become significant factors in sand sea accumulation, as evidenced by the 300 m of aeolian sand beneath the Murzuq Sand Sea in Libya (Glennie 1970). In contrast to the considerable lengths of time for sand sea accumulation envisaged by Wilson and Glennie, Ahlbrandt et al. (1983) argue that the 9-24 m thickness of sand in the Nebraska Sand Hills was deposited during the period 6000-8000 BP.

Evidence for the age of arid conditions in the world's major deserts, summarised by Street (1981), suggests that aridity developed from the late Miocene and early Pliocene (3-7 Ma ago) onwards. It is probable that most sand seas in existence today have developed during the late Cenozoic. As a result, climatic changes during the Quaternary have played an important role in determining the nature and rate of their development. The legacy of past wind regimes, in the form of dune patterns out of equilibrium with modern winds has been frequently cited as an example of the effects of such climatic changes (Glennie 1970, 1983; Besler 1980, 1982). More important as evidence for climatic change are the interdune pan, marsh and shallow lacustrine deposits which have been reported from sand seas in the Sahara (Rognon 1982); Saudi Arabia (Whitney et al. 1983) and the Namib (Teller and Lancaster 1985, 1986a). These deposits reflect periods of much wetter conditions and possibly stabilisation of nearby dunes. Bounding surfaces of regional extent may result from dune stabilisation by vegetation and pedogenic processes in periods of increased rainfall (Talbot 1985).

7.1 MODELS FOR SAND SEA ACCUMULATION

Many earlier workers (e.g. Aufrere 1928; Capot Rey 1945, 1970) noted the apparently close correspondence between the location of sand seas and low lying areas in desert regions and argued that the dunes of the sand seas were formed by the reworking of underlying fluvial and lacustrine deposits. A similar model was adopted by Besler (1980, 1984) to explain the formation of the Namib Sand Sea. She suggested that the Namib Sand Sea originated by the aeolian reworking of sand eroded from the Tsondab Sandstone Formation in the region of the Great Escarpment and deposited as a series of coalescing alluvial fans to the west, in a period coeval with the 'High Wurm'.

Compilations of wind data together with information from aerial photographs and satellite imagery (Wilson 1971; Mainguet 1972, 1977, 1978, 1983; Fryberger and Ahlbrandt 1979) have shown that long distance transport of sand by the wind occurs in the Saharan, Saudi Arabian and possibly Australian deserts. Deposition of sand and the accumulation of sand seas along these sand transport paths occurs wherever sand transport rates are reduced as a result of changes in climate or topography.

Wilson (1971) recognised three situations in which sand seas might develop. Flow centre ergs form where sand flow lines from all directions converge; saddle ergs occur where sand flows from two opposite directions converge, with a gap between two sand seas at the divergent peak of the sand flow lines. Flow crossed ergs are those in which sand flow lines cross the sand sea, with local convergence and deceleration giving rise to deposition. Wilson (1971) went on to argue deductively that four main factors influence the growth of sand seas with bedforms: (i) Deposition takes place even when sand flow is not fully saturated (actual sand flow rate is still less than potential). (ii) Sand sea growth is initially by lateral extension rather than vertical accretion. (iii) Sand sea shape and actual sand flow rates and deposition patterns are time dependent until bedform growth is complete. (iv) Bedform type influences sand sea development: sand seas which are dominated by 'sand passing' dunes such as linear dunes extend by bedform extension at their downwind margins and grow slowly whereas sand seas with a high proportion of 'sand trapping' bedforms (crescentic and star dunes) grow by both bedform migration at the downwind margin and bedform growth at the upwind margin. Most sand seas are a combination of 'sand passing' and 'sand trapping' bedforms.

Mader and Yardley (1985) have distinguished three main depositional mechanisms operating in aeolian systems: (i) Migration, which comprises downwind movement of dunes under constant wind directions. Deposition is by bedform climbing. (ii) Modification of dunes by winds from directions other than those which produced bedform migration. (iii) Merging, in which dunes coalesce with negligable erosion, as migration is slowed by topographic barriers or regional climatic changes.

Many sand seas accumulate where changes in wind regimes give rise to decelerating or converging sand flows. Reduction of sand transport rates may occur as a result of topographic barriers, which extend across and block sand flow lines. Sand seas begin to accumulate on the upwind lower slopes of such barriers and extend towards oncoming sand transport (Mainguet 1984). The Grand Erg Oriental in the Sahara and Great Sand Dunes, Colorado are two examples of the operation of such a mechanism (Fryberger and Ahlbrandt 1979).

Alternatively, decelerating winds and potential sand transport rates may occur as a result of regional changes in wind strength and directional variability, caused by regional changes in circulation patterns. The occurrence of sand seas in areas of low total or net sand transport has been noted for the western and southern Sahara by Fryberger and Ahlbrandt (1979) and Mainguet (1984). Convergence of winds and sand flows in the lee of major or minor topographic disturbances (Mainguet and Callot 1978) or areas of low elevation (Wilson 1971) enforces deposition by decreasing actual sand flow rates. Thus most sand seas appear to occur in areas where potential sand transport rates are lower and more variable in direction, compared to adjacent areas without sand sea development (Fryberger and Ahlbrandt 1979).

Wilson (1971) and Mainguet and Chemin (1983) have suggested that many sand seas are crossed by sand flows and that the same winds which transport sand to the sand sea can also remove it at the downwind end. These are 'flow crossed' sand seas in the terminology of Wilson (1971). Examples of this type of sand sea are the Makteir and other sand seas in Mauretania and the Nafud of Saudi Arabia (Fryberger and Ahlbrandt 1979).

7.2 MECHANISMS FOR SAND SEA ACCUMULATION

The lateral extent and thickness of aeolian sandstones suggests that they accumulated in ancient sand seas which persisted for long periods of time (McKee 1979a; Glennie 1970, 1983; Kocurek 1981a; Mader and Yardley 1985; Porter 1986). Typically, aeolian sandstones consist of sets of large scale cross strata separated by a hierarchy of bounding surfaces (Brookfield 1977). Most aeolian cross strata in the rock record have been ascribed to the migration of transverse dunes (e.g. Shotton 1937; Kocurek 1981a, b; Blakey and Middleton 1983) but linear dunes have been recognised by Tanner (1965), Glennie (1972) and Steele (1983).

Development of cross stratified beds is the product of bedforms migrating under conditions of deposition, or bedform climbing (Allen 1970; Rubin and Hunter 1982). Sand sea accumulation must therefore involve bedform climbing, which in turn requires dune migration, which also results in the development of a sequence of aeolian facies (Porter 1986). As they migrate, dunes may be modified by changing wind regimes or merge together (Mader and Yardley 1985).

Bedform climbing occurs principally as a result of decreases in sediment transport rates in the direction of dune migration (Rubin and Hunter 1982). Data on the wind environments of modern sand seas (Fryberger and Ahlbrandt 1979) strongly suggests that they occcur in areas of decreasing total and net sand transport and this is probably a primary cause of bedform climbing in most sand seas. Changes in transport rates over individual dunes also give rise to bedform climbing, both of ripples on dunes and superimposed dunes on compound and complex dunes. This gives rise to a threefold hierarchy of bedform climbing in a sand sea: at the simple dune, compound or complex dune or draa and sand sea scales, which corresponds to Brookfield's hierarchy of bounding surfaces. Accumulation of the sand sea involves both dune growth in addition to growth of the sediment body as a whole.

7.2.1 Dune growth

As suggested by Brookfield (1977), Kocurek (1981b), Ross (1983), Rubin and Hunter (1982, 1985), and Mader and Yardley (1985) complex and compound dunes (draa) probably grow mostly by a combination of merging and modification as a result of the movement of superimposed bedforms across macrodunes. Modern complex and compound dunes can be observed to develop by merging, or shingling, as faster moving smaller dunes accrete onto larger, slowly moving examples as a result of downcurrent decreases in migration rates (Howard et al. 1978). This may be the product of topographic barriers, as in the case of the Great Sand Dunes, Colorado (Andrews 1981; McKee 1983) or regional climatic change (McKee 1982). In the Namib Sand Sea, there appears to be a S-N decrease in wind strength and potential sandflow rates in coastal areas leading to the merging of crescentic dunes between Conception Bay and Sandwich Harbour. In the Gran Desierto, Mexico (Lancaster et al. 1987) crescentic dunes migrate towards and accrete onto large star and reversing dunes.

Climbing of superimposed bedforms on the lee slopes of macro dunes or draa is a major mechanism for dune growth. Its significance has been recognised by Brookfield (1977), Hunter and Rubin (1983) and Rubin and Hunter (1985). In the Namib Sand Sea superimposed barchanoid dunes on the eastern flanks of complex linear dunes and star dunes migrate from areas of higher to lower wind velocities and sand flow rates, as documented in Chapter 5. Following Rubin and Hunter (1985), the angle of climb of a dune is given by

$$\tan \zeta = \frac{\text{net rate of deposition } (V_z)}{\text{dune migration rate } (V_x)} \tag{1}$$

The dune migration rate is given by

$$V_x = \frac{2q \sin \alpha}{H} \tag{2}$$

where q is the mean bulk volume sediment transport rate, H is dune height and α is the angle between the dune trend and the resultant direction of sand transport (90°) for a transverse dune.

The rate of deposition (V_z) can be evaluated in terms of the sediment continuity equation (Middleton and Southard 1978) such that

$$V_z = -\left(\frac{\delta q_x}{\delta x} + \frac{\delta q_y}{\delta y}\right) \tag{3}$$

where q_x and q_y are the bulk volume sediment transport rates in the x and y directions respectively. As sand transport may cross a dune obliquely

$$V_z = \left(\frac{\sin \alpha \, \delta q_x}{\delta x} + \frac{\cos \alpha \, \delta q_y}{\delta y}\right) \tag{4}$$

where x and y are distances in the across dune and dune parallel directions and α is the angle between the dune and the direction of sand transport. X and Y can be regarded as the dimensions of the depositional area (D) over which transport rates decrease. This could be larger than the dune if sand leaves the dune or smaller than or equal to the dune if all sand remains on it. It is possible to simplify Equation (4) such that

$$V_z = \frac{2q}{D} \tag{5}$$

where D represents the distance over which sand transport rates decrease from their initial values on entering the depositional area to a final value. By substituting Equations (5) and (2) in Equation (1)

$$\tan \zeta = H (D' \sin \alpha)^{-1} \qquad\qquad (6)$$

where $D' = D/H$ and H is the average height of the dunes in the area.

In the case of superimposed barchanoid dunes on the east flank of linear dunes, measurements of wind velocities suggest that potential sand transport rates decrease to zero in all but the strongest winds over a distance of some 300 m, or the width of zone of east flank dunes. The average height of these dunes is 2.5 m and as they are transverse to S-SSW winds α is 90°. The average angle of climb (ξ) from Equation 6 is 0.48°, with a range between 0.29 and 1.29°. This compares with calculated angles of bedform climb of superimposed aeolian bedforms of 0.3° (Rubin and Hunter 1982) and measured angles from the rock record of 1.5° (Kocurek 1981b). The low angles of climb suggest that accumulation of these dunes is a slow process.

7.2.2 Sand sea growth

Using the expressions for the angle of bedform climb outlined above, it is possible to evaluate angles of bedform climb and deposition rates for the Namib Sand Sea as a whole, as well as for parts of its area.

If it is assumed that no sand leaves the Namib Sand Sea on its northern and eastern margins, the length of the depositional area, D is 400 km. Average dune height in the sand sea is 80 m and resultant sand flows cross its north-south axis at an average angle of 25°. From Equation (6) above, the calculated angle of bedform climb will be 0.003°. This compares with modal angles of bedform climb of 0.3° for crescentic dunes in the Entrada Sandstone of the western USA (Kocurek 1981b). In the northern part of the sand sea, where dunes are an average of 100 m in height, the depositional area is 150 km in width and crossed by sand flow at an average angle of 40°, calculated angles of bedform climb will be in the order of 0.06°. Angles of bedform climb for the Namib Sand Sea are therefore very low and suggest that very little sand accumulation below the bedforms has taken place. This is consistent with the widespread occurrence of sub-dune surfaces exposed in interdune areas. Angles of bedform climb on the large linear dunes appear to be greater than that of the dunes themselves, suggesting that they may be still growing.

Locally, deposition rates may be much higher, as sand transport rates in some areas appear to decline considerably over short distances. If data on winds, potential sand flow and dune migration rates is available for small areas of sand seas, then it is possible to infer areas in which deposition is rapid. One such area lies northeast of Luderitz at the southern margin of the sand sea where potential sand transport rates decline by a factor of 4 in a distance of some 40 km. Calculated deposition rates in this area are 0.25 m.yr^{-1} and 5 m high crescentic dunes migrate at 40 m.yr^{-1}, giving an angle of bedform climb of 0.03°. On the northwestern margin of the sand sea east of Sandwich Harbour, rates of potential sand flow decline by a factor of 10 in a distance of 20 km. In this area, calculated deposition rates are some 0.11 m.yr^{-1} and the migration rate of 10 m high crescentic dunes averages 4 m.yr^{-1}, giving an angle of bedform climb of 0.16°.

Both examples above are in areas in which shingling or merging of crescentic dunes is evident on air photographs. At the present time, it is probable that the inland margin of the area of crescentic dunes is the locus of most active deposition in the sand sea. Elsewhere, deposition rates appear to be very low. Currently, very slow migration and areal expansion appear to dominate the accumulation of the Namib Sand Sea.

The expressions for angles of bedform climb developed by Rubin and Hunter (1982, 1985) make it possible to evaluate the relationships between deposition rates and the dimensions of the sand sea. Sand seas which are crossed by sand transport paths will accumulate more slowly than those in which all sand is deposited. Sand seas aligned near parallel to sand transport directions will accumulate more slowly than those crossed by sand flows which are oblique or transverse to their major axis. This suggests that rates of sand sea accumulation will be controlled by five main factors: (i) The rate of sand transport into and out of the sand sea (sand supply). (ii) Regional gradients of changes in sand transport rates (deposition rates). (iii) Relationships between resultant sand transport directions and sand sea orientations. (iv) Rates of dune migration, which are in turn related to dune size and sand transport rates. (v) The size of the sand sea in relation to sand supply and deposition rates.

7.3 DEVELOPMENT OF THE NAMIB SAND SEA

7.3.1 *Sources of sand*

Two major hypotheses concerning the source of sands for the Namib Sand Sea have been put forward. Early investigators mostly working in the southern parts of the Namib (Wilmer 1883; Rand 1920; Harger 1914; Kaiser 1926) observed the extensive deflated and wind eroded plains of the Sperrgebiet and concluded that the dunes to the north had developed from sands derived from the Namib between Luderitz and the Orange River. Later workers, such as Gevers (1936), Logan (1960) and Hallam (1964) suggested that, in addition, sediments derived ultimately from the Orange River provided the sand for the Namib Sand Sea. Such a model was developed by Rogers (1977) who showed the close similarities between the texture and mineralogy of shallow shelf, beach and dune sands in the southern Namib (Fig. 83).

Besler and Marker (1979) drew attention to the existence of areally extensive semi consolidated red-brown sandstones of the Tsondab Sandstone Formation which are up to 100 m thick. They underlie much of the present area of the Namib Sand Sea and also outcrop in areas between its eastern margin and the base of the escarpment. Besler and Marker stated that the Tsondab Sandstone constituted the source of the dune sands. Besler (1980, 1984) further argued that changes in dune sand colour, grain roundness and grain size from east to west reflected a pattern in which sands were transported from east to west by fluvial action and deposited in a series of alluvial fans which were later reworked by aeolian processes into the dominant S-N linear dunes.

Lancaster and Ollier (1983) considered possible sources of sand in fluvial and shallow marine sediments as well as the Tsondab Sandstone Formation and older bedrock. Kuiseb River sediments were found to be distinctly different from dune sands with the light fraction consisting of over 90% quartz and the remainder feldspar. The heavy mineral assemblage of Kuiseb River sediments is dominated by opaque minerals, biotite (20-

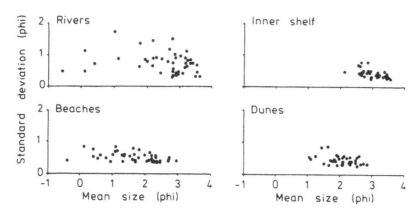

Figure 83. Comparative relationships between textural parameters of fluvial, beach, shelf and dune sands from the southern Namib hinterland and coastal zone. After Rogers (1977).

40%) and hornblende (< 20%), derived from Kuiseb Formation schists.

Most rivers draining from the escarpment to the east of the sand sea have small catchment areas and flow so infrequently that they do not constitute significant sources of sand sized material for dune formation. However, locally, along the lower courses of the Tsondab and Tsauchab rivers, small dunes and extensive sand sheets of grey sand, often including a substantial content of dolomite and limestone fragments, have formed by deflation of these river sediments.

The mineralogy of the light fraction of the Tsondab Sandstone Formation is very similar to that of the dune sands, consisting mostly of quartz (90%). The heavy minerals are dominated by opaques (> 80%) with garnet, clinopyroxene, amphibole, epidote and rare zircon, tourmaline and rutile. Locally, especially along the eastern margin of the sand sea it may be an important source for dune sands. In this area outcrops of Tsondab Sandstone Formation are widespread and weather rapidly to produce sand which is easily transported by wind. Many outcrops generate small shadow or falling dunes on their downwind margins. It is possible that a source in the Tsondab Sandstone Formation may explain the presence of very well sorted, but relatively coarse and leptokurtic sands in many eastern areas of the sand sea, especially in the area close to Sossus Vlei.However, the Tsondab Sandstone Formation does not appear to be a major source for sand of the sand sea.

Data on the mineralogy of the shelf sands (Ahmed 1968) shows that samples from the inner shelf between the Olifants and Orange rivers contain pyroxene, amphiboles, magnetite, garnet and zircon, with minor rutile, epidote, tourmaline, illmanite, kyanite and staurolite. Sediments in the fine sand range from north of the Orange River studied by O'Shea (1971) contain garnet, illmenite, rutile, tourmaline, staurolite, pyroxene and amphibole. Both these groups of samples have a composition very like that of the sands of the Namib Sand Sea, with dominant clinopyroxene, garnet and opaques (especially magnetite and illmenite) and traces of other minerals. Garnets, biotite and opaques are common in Damara Sequence rocks north of the Kuiseb River and in the Conception-Meob area, whilst clinopyroxenes are commonest in rocks of the Namaqualand Meta-morphic province in the Sperrgebiet and along the Orange River.

138

The importance of the Orange River and its sediments as a source for the sand in the Namib Sand Sea is emphasised by Rogers (1977, 1979). The river constitutes a major sediment source lying upwind of the sand sea. The water and sediment discharge of the Orange River are highly variable. Mean annual discharge is 10.30×10^7 m^3 with a variability of 49.4%. Mean annual sediment discharge varies between 8 and 326×10^6 tonnes with a mean of 60×10^6 tonnes, or 16.5×10^7 m^3 (Dingle et al. 1983), equivalent to a sediment yield of 14.98 m^3 km^{-2} per annum. This constitutes some 27% of the total sediment discharge from the southern African subcontinent (Dingle et al. 1983). Much of the sediment reaching the coast is fine to very fine sand, derived from Karoo Supergroup sediments. According to Rogers (1977) some 25-55×10^6 tonnes of sand sized sediment reach the sea via the Orange River each year. There is evidence (Rooseboom and Harmse 1979) to suggest that this rate is inflated by recent soil erosion and considerably in excess of the values predicted by the offshore sedimentary record, which indicate that sediment yields over the Neogene averaged 0.33 m^3 km^{-2} per annum (Dingle and Hendey 1984), a value comparable with many other desert rivers (Milliman and Meade 1983).

The Orange River delta is dominated by waves (Rogers 1977) and sand and mud are flushed out by summer floods. Whilst most mud is carried south by near shore currents, vigorous longshore drift carries sand sized material northwards for 200 km at a rate of 11 $\times 10^3$ tonnes per year 60 km north of Oranjemund (Rogers 1977). The textural characteristics of nearshore shelf, beach and dune sands in this area are very similar (Fig. 83) (Rogers 1977). Princenbucht and Possession Islands and the major re-entrant of Elizabeth Bay funnel much of the material on-shore, where it is deflated from the wide flat dissipative beaches and moved inland in prominent barchan trains (Fig. 84) under the influence of strong and persistent southerly winds. The Elizabeth Bay- Kolmanskop sand stream has probably been active episodically since the mid-late Miocene, as evidenced by the widespread occurrence of northward dipping cross bedded aeolian sandstones of the Fiskus Beds of the Elizabeth Bay Formation in this area.

The pattern of grain size and sorting parameters in the sand sea, discussed in Chapter 4, indicates that sand has been transported from southern and western coastal source zones to zones of accumulation in central and northern parts of the sand sea. There is a consistent fining trend, an improvement in sorting and a decline in skewness values from south to north and west to east across the sand sea. These patterns are best explained by a progressive loss of coarse grains in the direction of sand transport from sources to sinks. Local sources of sand, perhaps in the Tsondab Sandstone Formation, are indicated by the areas of relatively coarse, but very well sorted sands with a distinctive kurtosis value, along the eastern margins of the sand sea.

7.3.2 *Movement of sand into the sand sea*

Winds in the coastal areas of the Namib are predominantly S-SSW, or onshore. Throughout most of the sand sea the annual resultant direction of sand transport is towards the north and northeast. As discussed in Chapters 2 and 5 the strength and persistence of winds and thus the rate of sand transport decreases in the region of the Namib Sand Sea from south to north and from the coast inland. In addition, there is a parallel increase in the directional variability of winds and sand transport rates.

Consequently, sand is transported from southern and western coastal areas, which have

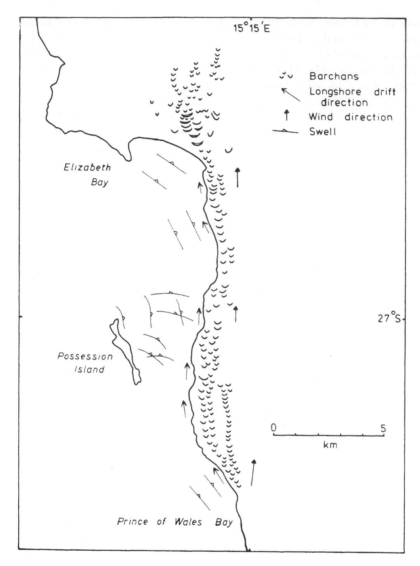

Figure 84. Relationships between coastal morphology and barchan dune initiation, Elizabeth Bay area. After Rogers (1977).

been identified as source zones, towards central, northern and eastern parts of the sand sea (Fig. 85). As sand transport rates decrease, the requirements of continuity dictate that deposition occurs. Further, sand is being transported into areas where winds are opposed in direction. Both total and net (resultant) sand transport rates in these areas are low. Consequently once sand reaches such areas, it will tend to be deposited.

Transport of sand into the central areas of the sand sea is by two related mechanisms: saltation and traction (bedload transport) and dune migration and extension (bedform transport). The amount of sand transport by dunes appears to be small by comparison

140

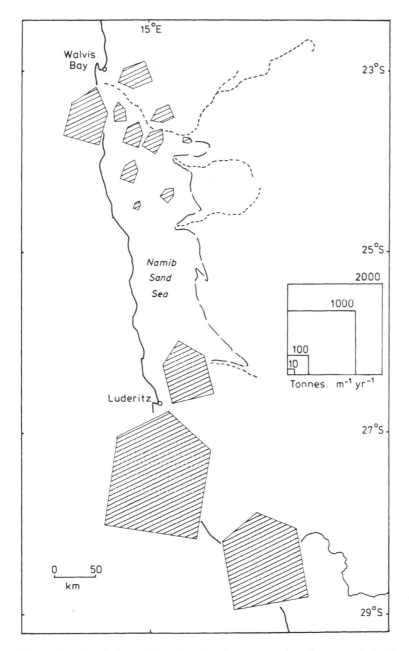

Figure 85. Magnitude and direction of resultant potential sand transport in the Namib Sand Sea.

with bedload transport. In areas of strong winds, such as the southern Namib, small crescentic dunes migrate rapidly and bedform transport of sand is probably very effective. Even here, bedload transport of sand through intervening rocky desert areas is much greater than that by bedforms. In a topographically similar area of the western Sahara, Samthein and Walger (1974) estimate that sand transport by saltation is 50-100 times greater than transport by migrating barchans.

In areas of linear dunes lee side flow diversion gives rise to sand transport along dunes leading to their extension, but especially in periods of strong SSW-SW and E-NE winds, sand crosses from dune to dune across interdune areas. The rates of sand transport in this manner far exceed sand transported by the migration of the dunes, which, as indicated in Chapter 5, averages less that 1 m per year. In southern areas of the sand sea dune parallel winds are common, such that sand transport along interdunes is effective.

7.3.3 *The pattern of deposition in the sand sea*

The pattern of dune size and spacing in the Namib Sand Sea is such that large complex linear and star dunes occur in central and northern areas of the sand sea. These large dunes represent a net accumulation of sand. Towards the margins of the sand sea progressively smaller compound and simple linear and crescentic dunes occur and dunes are generally low in the southern parts of the sand sea. From the relationships between dune height and equivalent sand thickness (Fig. 81), it is possible to estimate equivalent sand thickness for the sand sea as a whole (Fig. 86). This is a minimum figure for many parts of the sand sea, as it does not take into account sand which has accumulated below the bedforms and in interdune areas. Figure 86 shows that the accumulated thickness of sand over most of the central parts of the sand sea exceeds 20 m and is more than 30 m north and northwest of Sossus Vlei. Apart from small areas of star dunes, equivalent sand thickness in the southern parts of the sand sea is less than 10 m. The pattern of deposition indicated by Figure 86 is further evidence for the accumulation of the Namib Sand Sea by transport of sand from southern and western source zones to accumulate in central and northern parts of the sand sea.

7.3.4 *Effects of Quaternary climatic fluctuations*

Quaternary climatic fluctuations in the Namib appear to have been of a low amplitude and superimposed upon a trend of increasing aridity (Ward et al. 1983; Lancaster 1984b). This is evidenced by the surface survival of calcrete paleosols (Yaalon and Ward 1982), calcareous lacustrine deposits (Selby et al. 1979) and by marine sediments which record a dominance of aeolian inputs in the past 200,000 years (Lancaster 1984b). On the basis of pollens in DSDP cores from the Walvis Ridge, Van Zinderen Bakker (1984a) suggests that the vegetation of the region has not changed significantly since the Pliocene. A trend to increasing aridity is suggested by the decline in clast size in the deposits of most Namib rivers, which implies that river discharge and competence have decreased during the Pleistocene (Korn and Martin 1957; Lancaster 1984b).

Rognon (1982), referring to Saharan sand seas, has pointed to the contrasting effects of periods of wetter and drier climates on sand seas. In periods of increased rainfall, the high infiltration capacity and porosity of dune sands favours the growth and persistence of

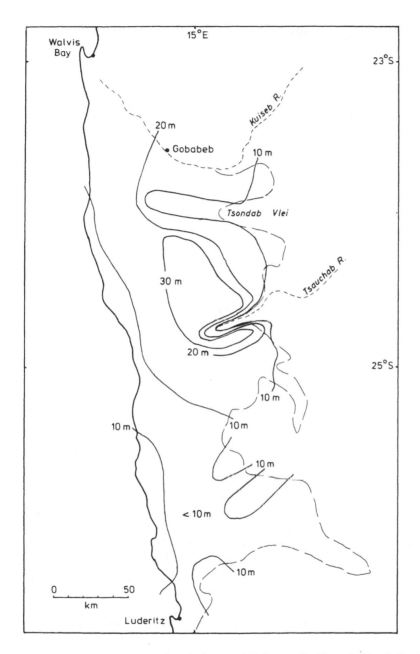

Figure 86. Spatial variation of equivalent sand thickness of bedforms in Namib Sand Sea.

143

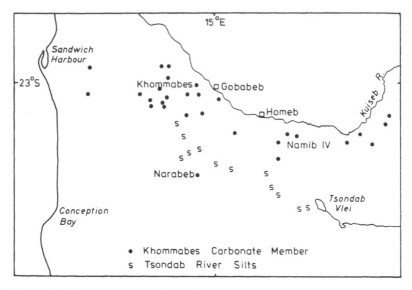

Figure 87. Sites of exposures of interdune pond and marsh deposits in the northern part of the Namib Sand Sea. After Ward (1984).

vegetation, which may lead to partial or complete stabilisation of dunes. If the periods of increased rainfall are of sufficient duration, water tables will rise, leading to the accumulation of pond and marsh deposits in interdunes. In periods of intense rainfall, runoff may erode and degrade dunes (Talbot 1985). In contrast periods of aridity will give rise to a very unfavourable biotic environment within the sand sea, with low water tables and active sand movement retarding vegetation growth.

In the Namib Sand Sea periods of increased moisture availability, and probably also rainfall, are recorded by scattered interdune pond and marsh deposits (Fig. 87) which are particularly common in the northern part of the sand sea (Ward 1984). Archaeological evidence in the form of artefact scatters points to a number of periods when the sand sea presented, as least seasonally, a favourable habitat for man. Middle Stone Age artefacts are found widely throughout the sand sea and in the most arid parts of the desert, whilst Later Stone Age artefacts tend to be concentrated near to present seasonal or permanent water sources (Korn and Martin 1957).

The earliest well dated period of higher rainfall in the Namib Sand Sea is documented by shallow lacustrine pan carbonates at Namib IV, in the northeastern part of the sand sea. The deposits contain bones of *Elephas recki* and an associated Acheulian artefact industry, suggesting a Mid Pleistocene age of 400-700,000 BP (Shackley 1980). Inferred environmental conditions are suggestive of a semi arid climate and savanna vegetation, although this interpretation has been questioned by Ward et al. (1983).

During the Late Pleistocene, pan and shallow lacustrine carbonates and associated reed beds from a variety of sites have radiocarbon dates clustering around 30-28,000 and 24-23,000 BP (Vogel and Visser 1981; Deacon et al. 1984; Teller and Lancaster 1985, 1986a). During this period water tables at Khommabes and elsewhere near Gobabeb

144

were higher, leading to the growth of reeds and the development of shallow lakes in interdune areas. Further south, at Narabeb, calcareous muds accumulated in the period 26-20,000 BP at the end point of the Tsondab River, fed by increased river flow from the highlands to the east through a valley which was not blocked by dunes until after 12,000 BP (Seely and Sandelowsky 1974; Teller and Lancaster 1986b). On the southern margins of the sand sea, at Koichab Pan, reed beds grew in conditions of higher runoff or groundwater discharge immediately prior to 23,000 BP. However, analyses of pollen from silts deposited at Sossus Vlei (Van Zinderen Bakker 1984b) suggest that, although there was more runoff and river flow at times, the increase in rainfall was not sufficient to significantly modify the composition of the local vegetation.

During these periods, when it is probable that rainfall within the area of the the Namib Sand Sea was higher than today, partial or complete stabilisation of the dunes by vegetation may have occurred. Under present conditions, 50-100 mm of rain a year is required to generate the growth of perennial grasses on the dunes such that only crestal areas are active. Rainfall increases of at least 2-3 times present amounts are implied for periods of dune stabilisation.

Evidence for climatic conditions as dry as, or drier than the present is usually lacking from desert sand seas, as dateable materials are usually deposited under conditions of increased humidity. This is the case in the Namib Sand Sea, but absence of evidence indicating increased moisture availability has been taken to indicate aridity in the period between 20,000 and 16,000 BP by Deacon et al. (1984) and Lancaster (1984b). During the period coeval with the Last Glacial Maximum, conditions of increased wind velocity and the extension of a wind regime with more southerly winds, such as occurs today south of Luderitz, to the whole sand sea were inferred by Besler (1980) from the dune alignment pattern. However, no independent evidence is available to support such a hypothesis. Lancaster (1981b) studied the pattern of alignments of fixed dune systems throughout the subcontinent and concluded that, in the late Pleistocene, circulation patterns were displaced by a maximum of 2° of latitude, or 200 km. However, in the latest period of dune formation, during the period coeval with the Last Glacial Maximum, there is no evidence for latitudinal shifts in circulation patterns. Rather, increased dune activity was the product of higher wind velocities generated by intersified anticyclonic circulations. Similar conclusions have been reached for the Sahara by Mainguet and Canon (1976) and Talbot (1980), and for Australia by Wasson (1983b). Estimates of the increases in wind velocities during this period by Newell et al. (1982) show that wind velocities in the latitude of the Namib were 117% of present. This would have increased sand movements throughout the sand by 2.35-6 times, with the effect being most marked in lower energy wind regimes, such as those which predominate in central and northern parts of the sand sea.

7.3.5 *Effects of Quaternary sea level changes*

Because of the coastal location of the sand sea, eustatic changes in sea level during the Quaternary era may be expected to have had an important effect upon its accumulation. Rogers (1977) has mapped the widespread occurrence of quartz sand in shelf sediments off the Namib coast. In periods of sea level regression such sediments would have been a readily deflated source of sand for input to the sand sea. The association of regressions with glacial advances at high latitudes also means that such periods may well have been windier than at present.

During the period of the Last Glacial Maximum, 20-18,000 BP, sea levels off the Namib fell by 110 m (Rogers 1977) exposing up to a 20-30 km width of the shelf to wind action. In addition, changes in the orientation of shorelines created a major re-entrant south of Baker Bay (Rogers 1977). At a sea level of -20 to -40 m, a south west facing shoreline probably existed south of Meob Bay (Fig. 88) giving rise to conditions favourable for onshore movement of sand to feed the northern group of compound crescentic dunes, which at present lack a suitable source.

Periods of sea level rise, such as the Post Glacial or Flandrian transgression and the 4-6 m higher sea levels of the Last Interglacial may have liberated large amounts of sand by wave reworking of the margins of the sand sea and of dunes which existed in the area of what is now the inner shelf. By analogy with coastal dune areas on the southern coast of South Africa (Butzer and Helgren 1972), the Post Glacial transgression may have provided a short lived pulse of sand for input to the sand sea. However, transgressions may be expected to have reduced aeolian sand input to the sand sea.

Periods of lowered sea levels may therefore have been associated with extension of the area of the sand sea, whilst transgressions may have given rise to a concentration of sand within the existing sand body, leading to its vertical accretion. In addition, increased wind velocities during Glacials may have favoured rapid dune extension and migration, whilst weaker Interglacial winds, such as today, may have given rise to reduced dune movement, but perhaps more vertical growth of dunes by deposition. Periods of more humid climates in the region, although of low amplitude, provided a localised input to the sand sea in the form of increased fluvial sedimentation and contributed to the development of non aeolian interdune sediments.

7.4 THE AGE OF THE NAMIB SEA SEA

7.4.1. *Evidence from the Cenozoic geological record*

As discussed in Chapter 2, considerable controversy has surrounded the age of desertic conditions in the Namib. A Pleistocene to Recent age for the sand sea is assigned by a number of workers (Tankard and Rogers 1978; Besler 1980; Ollier 1977; Lancaster 1982), but Ward et al. (1983), following Korn and Martin (1957) have suggested that it may have had its origins in the Pliocene.

Aeolian sandstones of a number of ages have been described from the southern and central Namib. The most extensive is the Tsondab Sandstone Formation which has been assigned an Early-Middle Tertiary age by Ward et al. (1983), although it could have been deposited at any time before the end Miocene, on stratigraphic grounds. The sandstones of the Fiskus Beds of the Elizabeth Bay Formation are extensively exposed in the Grillental-Kolmanskop area of the southern Namib and overlie Early-Mid Tertiary clays which contain a fauna suggestive of savanna conditions and a sub humid climate (Corbett pers. comm.). The sandstones were deposited by transverse dunes in a southerly wind regime and represent the first unequivocal evidence of aridity in this area. Correlations between the Fiskus Beds and other aeolian sandstones in the region, especially the

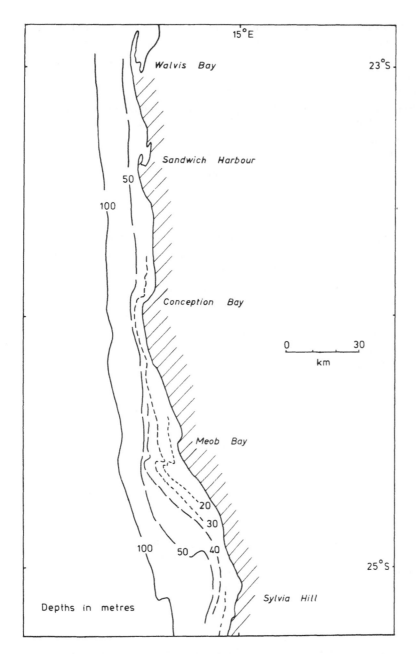

Figure 88. Relationships between bathymetry and modern coastline in the Conception-Meob area to show effects of sea level changes on coastal morphology and potential sites of dune initiation.

Tsondab Sandstone Formation, are at present uncertain, but it seems clear that there has been a history of aeolian activity in the region dating back at least to the Late Miocene. The Fiskus Beds and the Tsondab Sandstone Formation can be considered to be precursors of the present sand sea.

Evidence from the southern Namib (Corbett pers. comm.) and the offshore record (Siesser 1980) strongly suggests that the present hyper-aridity of the Namib originated in the Late Miocene and developed progressively thereafter. The Namib Sand Sea has therefore probably accumulated from the Pliocene onwards. Its initiation may have been promoted by the late Pliocene regression, which also led to the formation of aeolianites of the Alexandria Formation in the southern Cape Province of South Africa (Dingle et al. 1983).

Stages in the development of the sand sea are marked by large scale aeolian cross beds intercalated with fluvial gravels of the Oswater Conglomerate in the southern bank of the Kuiseb River. These deposits are assigned a Mid Pleistocene age by Ward (1982, 1984). Cross bedded aeolian sands are also associated with the mid Pleistocene pan carbonates at Namib IV (Ward et al. 1983) and it appears that large linear dunes were present on the southern banks of the Kuiseb River in the mid Pleistocene. Lenses of red aeolian sands are also common in the Homeb Silt Formation, which accumulated in the period 23,000-19,000 BP (Vogel 1982). However, Teller and Lancaster (1985) suggest that the first evidence of aeolian deposition in the Khommabes basin, some 5 km south of the Kuiseb River west of Gobabeb, is contained in small crossbedded dune sands which accumulated after 20,000 BP.

Although not conclusive, these observations seem to suggest that the northern margin of the sand sea may have fluctuated considerably, or been very irregular, during the Mid-Late Pleistocene. The large dune free area of the Tsondab Flats (Lancaster 1980) and the disturbance of the main S-N linear dunes to the west seems to suggest that the Tsondab River may have formed a barrier to northward extension of the sand sea until after 20,000 BP (Teller and Lancaster 1986b).

7.4.2 *Estimates based upon sand input*

The present Namib Sand Sea has a volume of some 1.02×10^{12} m³. Present day input of sand to the sand sea is primarily through the Elizabeth Bay-Kolmanskop corridor where current potential, and probably actual, sand transport is in the order of 2,000 tonnes. $m^{-1}.yr^{-1}$. This is equivalent to some 50 $m^3.m^{-1}.yr^{-1}$ or a bulk volume of 350,000 $m^3.yr^{-1}$ through the Kolmanskop corridor. Assuming that the Conception-Meob area and other source areas have made minor contributions, total input of sand to the sand sea is of the order of 400,000 $m^3.yr^{-1}$.

If it is assumed that present rates represent a mean figure for the Pleistocene, and all sand input stays in the sand sea, a minimum of 2.55 million years of sand transport would have been necessary for its formation. A minimum age for the Namib Sand Sea of some 2-3 Ma seems acceptable in the light of present day knowledge of the climatic and geomorphic history of the region.

REFERENCES

Ahlbrandt, T.S. 1974. The source of sand for the Killpecker sand dune field, southwestern Wyoming. *Sed.Geol.* 11: 39-57.

Ahlbrandt, T.S. 1979. Textural Parameters of Eolian Deposits. In E.D.McKee (ed.), *A Study of Global Sand Seas. US Geol.S.Prof.Paper* 1052: 21-51.

Ahlbrandt, T.S. & S.G.Fryberger 1980. Eolian deposits in the Nebraska Sand Hills. *US Geol.S.Prof.Paper* 1120-A.

Ahlbrandt, T.S. & S.G.Fryberger 1981. Sedimentary features and significance of interdune deposits. *S.E.P.M.Special Publication* 31: 293-314.

Ahlbrandt, T.S. & S.G.Fryberger 1982. Eolian Deposits. In P.A.Scholle and D.Spearing (eds.), *Sandstone Depositional Environments.* Amer.Ass.Petrol.Geol.Memoirs 31: 11-48.

Ahlbrandt, T.S., J.B.Swinehart & D.G.Maroney 1983. The Dynamic Holocene dunefields of the Great Plains and Rocky Mountain Basins, U.S.A. In M.E.Brookfield and T.S.Ahlbrandt (eds.), *Eolian Sediments and Processes*, Developments in Sedimentology 38. Amsterdam: Elsevier.

Ahmed, A.A.M. 1968. Geochemical and mineralogical studies of sediments from the South-West African shelf. Unpublished M.Phil. Thesis: London University.

Alimen, M.H. 1953. Variations granulométriques et morphoscopiques du sable le long de profiles dunaires au Sahara occidental. *Actions Eoliennes, Centre Nationale de Recherches Scientifiques, Colloques Internationaux* 35: 219-235.

Alimen, M.H., M.Buron & J.Chavaillon 1958. Caractéres granulométriques de quelques dunes d'ergs du Sahara nord-occidental. *Académie des Sciences, Paris*, C.R. 247: 1758-1761.

Allchin, B., A.S.Goudie & K.T.M.Hegde 1978. *The Prehistory and Paleogeography of the Great Indian Desert.* London: Academic Press.

Allen, J.R.L. 1968. *Current Ripples.* Amsterdam: Elsevier.

Allen, J.R.L. 1970. *Physical Processes of Sedimentation.* London: Allen and Unwin.

Allen, J.R.L. 1974. Reaction, relaxation and lag in natural sedimentary systems: general principles, examples and lessons. *Earth Sci.Rev.* 10: 263-342.

Allen, J.R.L. 1980. Sand waves: a model of their origin and internal structure. *Sed.Geol.* 26: 281-328.

Andrews, S. 1981. Sedimentology of the Great Sand Dunes, Colorado. *S.E.P.M.Special Publication* 31: 279-291.

Angell, J.K., D.H.Pack & C.R.Dickson 1968. A Langrangian study of helical circulations in the planetary boundary layer. *J.Atmos.Sci.* 25: 707-717.

Ash, J.E. & R.J.Wasson 1983. Vegetation and sand mobility in the Australian desert dunefield. *Z.Geomorph.Sup.Bd.* 45: 7-25.

Aufrere, L. 1928. L'orientation des dunes et la direction des vents. *Académie des Sciences, Paris, C.R.* 187: 833-835.

Bagnold, R.A. 1941. *The Physics of Blown Sand and Desert Dunes.* London: Chapman and Hall.

Bagnold, R.A. 1953. Forme des dunes de sable et regime des vents. *Actions Eoliennes, Centre National de Recherches Scientifiques, Colloques Internationaux* 35: 23-32.

Bagnold, R.A. 1954. Experiments on a gravity free dispersion of large solid spheres in a Newtonian fluid under shear. *Proc.Roy.Soc.* A 225: 49-63.

Barnard, W.S. 1973. Duinformasies in die Sentrale Namib. *Tegnikon* 5: 2-13.

Barnard, W.S. 1975. Geomorphologiese Processe en die mens: die geval an die Kuisebdelta South West Africa. *Acta Geogr.* 2: 20- 43.

Beheiry, S.A. 1967. Sandforms in the Coachella Valley, southern California. *Ann.Assoc.Amer.Geogr.* 57: 25-48.

Bellair, P. 1953. Sables désertiques et morphologie éolienne. *Proceedings 19th International Geological Congress*, Algiers, 1952, 7: 113-118.

Besler, H.1972. Klimaverhältnisse und klimageomorphologische Zonierung der Zentralen Namib. *Stuttgarter Geographische Studien*, Band 83.

Besler, H. 1975. Messungen zur Mobilität von Dünensanden am nord-Rand der Dünen-Namib. *Würzburger Geogr.Arb.* H43: 135-147.

Besler, H. 1976. Wasserüberformte dünen als Glied in der Landschaftgenese der Namib. *Basler Afrika Bibliographen* 15: 83- 106.

Besler, H. 1977. Untersuchungen in den Dünen Namib (Sudwesafrika). *J.South West Africa Sci.Soc.* 31: 33-64.

Besler, H. 1980. Die Dünen-Namib: Entstehung und Dynamik eines Ergs. *Stuttgarter Geographische Studien* 96.

Besler, H. 1982. The north-eastern Rub'al Khali within the borders of the United Arab Emirates. *Z.Geomorph.* 26: 495-504.

Besler, H. 1984. The Development of the Namib dune field according to sedimentological and geomorphological evidence. In J.C.Vogel (ed.), *Late Cainozoic Palaeoclimates of the Southern Hemisphere*. Rotterdam: Balkema.

Besler, H. & M.E.Marker 1979. Namib sandstone: a distinct lithological unit. *Trans.Geol.Soc.S.Afr.* 82: 155-160.

Binda, P.L. 1983. On the skewness of some eolian sands from Saudi Arabia. In M.E.Brookfield & T.S.Ahlbrandt (eds.), *Eolian Sediments and Processes*, Developments in Sedimentology 38. Amsterdam: Elsevier.

Blakey, R.C. & L.T.Middleton 1983. Permian shoreline eolian complex in central Arizona: dune changes in response to cyclic sea-level changes. In M.E.Brookfield and T.S.Ahlbrandt (eds.), *Eolian Sediments and Processes*, Developments in Sedimentology 38. Amsterdam: Elsevier.

Blandford, W.T. 1876. On the physical geography of the Great Indian Desert with especial reference to the existence of the sea in the Indus valley and on the origin and mode of formation of the sand hills. *J.A.S.B. (Calcutta)* 45: 86-103.

Bowen, A.J. & D.Lindley 1977. A wind-tunnel investigation of the wind speed and turbulence characteristics close to the ground over various escarpment shapes. *Boundary Layer Meteorol.* 12: 259- 271.

Bradley, E.F. 1980. An experimental study of the profiles of wind speed shearing stress and turbulence at the crest of a large hill. *Q.J.R.Meteo.S.* 106: 101-123.

Breed, C.S. 1977. Terrestrial analogs of the Hellespontus dunes, Mars. *Icarus* 30: 326-340.

Breed, C.S. & W.J.Breed 1979. Dunes and other windforms of central Australia (and a comparison with linear dunes on the Moenkopi Plateau, Arizona). In F.El-Baz & D.M.Warner (eds.), *Apollo-Soyuz Test Project vol 2: Earth Observations and Photography*. Washington DC: National Technical Information.

Breed, C.S. & T.Grow 1979. Morphology and distribution of dunes in sand seas observed by remote sensing. In E.D.McKee (ed.), *A Study of Global Sand Seas. US Geol.S.Prof.Paper* 1052: 253-304.

Breed, C.S., S.G.Fryberger, S.Andrews, C.McCauley, F.Lennartz, D.Geber & K.Horstman 1979. Regional studies of sand seas using LANDSAT (ERTS) imagery. In E.D.McKee (ed.), *A Study of Global Sand Seas. US Geol.S. Prof.Paper* 1052: 305-398.

Bremner, J.M. 1984. The coastline of Namibia. *Joint Geological Survey/ University of Cape Town Marine Geoscience Group Technical Report* 15: 200-206.

Brookfield, M.E. 1977. The origin of bounding surfaces in ancient aeolian sandstones. *Sedimentology* 24: 303-332.

Bull, W.B. 1975. Allometric change of landforms. *Geol.Soc.Amer. Bull.* 86: 1489-1498.

Butzer, K.W. & D.M.Helgren 1972. Late Cenozoic evolution of the Cape coast between Knysna and Cape St. Francis, South Africa. *Quat.Res.* 2: 143-169.

Cailleux, A. 1952. L'indice de émousée de grains de sable et grés. *Rev.de Geomorph.Dynam.* 2: 78-87.

Capot-Rey, R. 1945. Dry and humid morphology in the western erg. *Geogr.Rev.* 35: 391-407.

Capot-Rey, R. 1947. L'edeyen de Mourzouk. *Trav.Ins. de Recherches Sahariennes* 4: 67-109.

Capot-Rey, R. 1970. Remarques sur les ergs du Sahara. *Ann.Geog.* 79: 2-19.

Capot-Rey, R. & M.Gremion 1964. Remarques sur quelques sables Sahariennes. *Trav.Ins. de Recherches Sahariennes* 23: 153-163.

Carter, L.D. 1981. A Pleistocene sand sea on the Alaskan Arctic coastal plain. *Science* 211: 381-383.

Chavaillon, J. 1964. Les formations Quaternaires du Sahara nord-occidental. *C.N.R.S. Publ. Centre Recherches Zones Arides, Ser.Geol.* 5.

Chepil, W.S. & N.P.Woodruff 1963. The physics of wind erosion and its control. *Advan.Agron.* 15: 211-302.

Chorley, R.J. & B.A.Kennedy 1971. *Physical Geography*. London: Prentice Hall.

Clarke, R.H. & C.H.B.Priestley 1970. The asymmetry of Australian desert sand ridges. *Search* 1: 77-78.

Clos-Arceduc, A. 1967. La direction des dunes et ses rapports avec celle du vent. *Académie des Sciences, Paris*, C.R. 264 D: 1393-1396.

Cooke, R.U. & A.Warren 1973. *Geomorphology in Deserts*. London: Batsford.

Cooper, W.S. 1958. Coastal sand dunes of Oregon and Washington. *Geol.Soc.Amer.Memoir* 72.

Corvinus, G. & Q.B.Hendey 1978. A new Miocene vertebrate locality at Arriesdrift in Namibia (South West Africa). *Neues Jarbuch für Geologie und Paläontology*. Monatshefte 4: 193-205.

David, P.P. 1977. *Sand dune occurrences in Canada*. Department of Indian and Northern Affairs, National Parks Board Report, Ottawa.

Deacon, J., N.Lancaster & L.Scott 1984. Evidence for Late Quaternary climatic change in southern Africa: Summary of the Proceedings of the SASQUA Quaternary Workshop held in Johannesburg, September 1983. In J.C.Vogel (ed.), *Late Cainozoic Palaeoclimates of the Southern Hemisphere*. Rotterdam: Balkema.

Dingle, R.V. & Q.B.Hendey 1984. Late Mesozoic and Tertiary sediment supply to the eastern Cape basin (SE Atlantic) and palaeo-drainage systems in southwestern Africa. *Mar.Geol.* 56: 13- 26.

Dingle, R.V. & R.A.Scrutton 1974. Continental breakup and the development of post-Paleozoic sedimentary basins around southern Africa. *Geol.Soc.Amer.Bull.* 85: 1467-1474.

Dingle, R.V., W.G.Siesser & A.R.Newton 1983. *Mesozoic and Tertiary Geology of Southern Africa*. Rotterdam: Balkema.

Dresch, J. 1982. Sur la semi-aridite du Maghreb au Plio- Quaternaire.*Ann.Geog.* 431: 2-19.

El Baz, F. 1978. The meaning of desert color in earth orbital photographs. *Photogr.E.R.* 44: 71-75.

Embabi, N.S. 1982. Barchans of the Kharga Depression. In F.El Baz et al. (eds.), *Desert Landforms of Southwestern Egypt: a basis for comparison with Mars*. National Aeronautics and Space Administration, Contractor Rep. 3611: 141-155.

Endrody-Younga, S. 1982. Dispersion and translocation of dune specialist tenebrionids in the Namib area. *Cimbebasia* (A) 5: 257- 271.

Engel, P. 1981. Length of flow separation over dunes. *J. Hydra- A.S.C.E.* 107(MY10): 1133-1143.

Finkel, H.J. 1959. The barchans of southern Peru. *J.Geol.* 67: 614-647.

Folk, R.L. 1971. Longitudinal dunes of the northwestern edge of the Simpson desert, Northern Territory, Australia. 1: Geomorphology and grain size relationships. *Sedimentology*, 16: 5- 54.

Folk, R.L. 1976. Reddening of desert sands: Simpson Desert N.T. Australia. *J.Sed.Petrol.* 46: 604-615.

Folk, R.L. 1978. Angularity and silica coatings of Simpson Desert sand grains, Northern Territory, Australia. *J.Sed.Petrol.* 48: 611-624.

Folk, R.L. & W.C.Ward 1957. Brazos river bar: a study in the significance of grain size parameters. *J.Sed.Petrol.* 27: 3-26.

Fredsoe, J. 1982. Shape and dimension of stationary dunes in rivers. *J.Hydra-A.S.C.E.* 108 (HY8): 932-947.

Fryberger, S.G. 1979. Dune forms and wind regime. In E.D.McKee (ed.), *A Study of Global Sand Seas. US Geol.S.Prof.Paper* 1052: 137-140.

Fryberger, S.G. & T.S.Ahlbrandt 1979. Mechanisms for the formation of eolian sand seas. *Z.Geomorph.* 23: 440-460.

Fryberger, S.G., A.M.Al-Sari & T.J.Clisham 1983. Eolian dune, interdune, sand sheet and siliclastic sabkha sediments of an offshore prograding sand sea, Dhahran area, Saudi Arabia. *Am.Assoc.Pet.Geol.Bull.* 67(2): 280-312.

Fryberger, S.G., A.M.Al-Sari, T.J.Clisham, S.A.R.Rizoi & K.G.Al- Hinai 1984. Wind sedimentation in the Jafarah sand sea, Saudi Arabia. *Sedimentology* 31: 413-431.

Gardner, R. & K.Pye 1981. Nature, origin and paleoenvironmental significance of red coastal and desert dune sands. *Prog.Phys.Geogr.* 5: 514-534.

Gevers, T.W. 1936. The morphology of western Damaraland and the adjoining Namib Desert. *S.Afr.Geogr.J.* 19: 61-79.

Glennie, K.W. 1970. *Desert Sedimentary Environments*. Developments in Sedimentology 14. Amsterdam: Elsevier.

Glennie, K.W. 1972. Permian Rotliegendes of northwest Europe interpreted in light of modern desert sedimentation studies. *Am.Assoc.Pet.Geol.Bull.* 56: 1048-1071.

Glennie, K.W. 1983. Lower Permian Rotliegendes desert sedimentation in the North Sea area. In M.E.Brookfield and T.S.Ahlbrandt (eds.), *Eolian Sediments and Processes*, Developments in Sedimentology 38. Amsterdam: Elsevier.

Goudie, A. 1970. Notes on some major dune types in South Africa. *S.Afr.Geogr.J.* 52: 93-101.

Goudie, A. 1972. Climate, weathering, crust formation, dunes and fluvial features of the central Namib Desert, South West Africa. *Madoqua* 2: 54-62.

Goudie, A. 1983. The Arid Earth. In R.Gardner and H.Scoging (eds.) *Megageomorphology*. Oxford: Clarendon Press.

Goudie, A.S. & C.H.B.Sperling 1977. Long distance transport of foraminifera tests by wind in Thar desert, northwest India. *J.Sed.Petrol.* 47: 630-633.

Goudie, A.S. & A. Watson 1981. The shape of desert sand dune grains. *J.Arid Environments* 4: 185-190.

Goudie, A.S., R.U.Cooke & J.C.Doornkamp 1979. The formation of silt from quartz dune sand by salt weathering processes in deserts. *J.Arid Environments* 2: 105-112.

Gould, S.J. 1966. Allometry and size in ontogeny and phylogeny. *Cambridge Philos.Soc.Biol.Rev. 41: 587-640.*

Greeley, R. & J.D.Iversen 1985. *Wind as a Geological Processes on Earth, Mars, Venus and Titan*. Cambridge: Cambridge Univ.Press.

Hack, J.T. 1941. Dunes of the western Navajo country. *Geogr.Rev.* 31: 240-263.

Hallam, C.D. 1964. The geology of the coastal diamond deposits of southern Africa. In S.H.Houghton (ed.), *The Geology of Some Ore Deposits in South Africa*. Johannesburg: Geological Society of South Africa.

Hanna, S.R. 1969. The formation of longitudinal sand dunes by large helical eddies in the atmosphere. *J.Appl.Met.* 8: 874-883.

Harger, H.S. 1914. Some features associated with the denudation of the South African continent. *Proc.Geol.Soc.S.Afr.* 16: 22-41.

Harmse, J.T. 1982. Geomorphologically effective winds in the northern part of the Namib Sand Desert. *S.Afr.Geogr.* 10: 43-52.

Hastenrath, S.L. 1967. The barchans of the Arequipa Region, southern Peru. *Z.Geomorph.* 11: 300-331.

Holm, D.A. 1960. Desert geomorphology in the Arabian Peninsula. *Science* 132: 1369-1379.

Hovermann, J. 1978. Formen und Formung in der Pranamib (Flachen-Namib). *Z.Geomorph.Sup.Bd.* 30: 55-73.

Howard, A.D., J.B.Morton, M.Gad-el-Hak & D.Pierce 1978. Sand transport model of barchan dune equilibrium. *Sedimentology* 25: 307-338.

Hunter, R.E. 1977. Basic types of stratification in small eolian dunes. *Sedimentology* 24: 361-388.

Hunter, R.E. & D.M.Rubin 1983. Interpreting cyclic crossbedding, with an example from the Navajo Sandstone. In M.E.Brookfield & T.S.Ahlbrandt (eds.), *Eolian Sediments and Processes*, Developments in Sedimentology 38. Amsterdam: Elsevier.

Hunter, R.E., B.M.Richmond & T.R.Alpha 1983. Storm controlled oblique dunes of the Oregon coast. *Geol.Soc.Amer.Bull.* 94: 1450- 1465.

Huthnance, J.M. 1982. On one mechanism forming linear sand banks. *Est.Coast and Shelf Sci.* 14: 79-99.

Inman, P.L., G.C.Ewing & J.B.Corliss 1966. Coastal sand dunes of Guerrero Negro, Baja California, Mexico. *Geol.Soc.Amer.Bull.* 77: 787-802.

Jackson II, R.G. 1975. Hierarchial attributes and a unifying model of bed forms composed of cohesionless material and produced by shearing flow. *Geol.Soc.Amer.Bull.* 86: 1523-1533.

Jackson, P.S. 1977. A theory for flow over escarpments. In K.J.Eaton (ed.), *Wind Effects on Buildings and Structures*. Cambridge: Cambridge Univ.Press.

Jackson, P.S. & J.C.R.Hunt 1975. Turbulent wind flow over a low hill. *Q.J.R.Meteo.S.* 101: 929-955.

Kadar, L. 1934. A study of the sand sea in the Libyan Desert. *Geogr.J.* 83: 470-478.

Kaiser, E. 1926. *Die Diamantenwuste Südwes-Afrikas*. Berlin: Reimer.

Kennedy, J.F. 1969. The formation of sediment ripples, dunes and antidunes. *Ann.Rev.Fluid Mech.* 1: 147-169.

Kennett, J.P. 1980. Palaeoceanographic and biogeographic evolution of the Southern Ocean during the Cenozoic, and Cenozoic microfossil datums. *Palaeogeogr.Palaeoclimatol. Palaoecol.* 31:123-152.

King, D. 1960. The sand ridge deserts of South Australia and related aeolian landforms of Quaternary arid cycles. *Trans.R.Soc.S.Aust.* 79: 93-103.

Knott, P. 1979. The structure and pattern of dune forming winds. Unpubl. PhD Thesis: University of London.

Kocurek, G. 1981a. Erg reconstruction: the Entrada sandstone (Jurassic) of northern Utah and Colorado. *Paleogeogr.Paleoclimatol.Paleoecol.* 36: 125-153.

Kocurek, G. 1981b. Significance of interdune deposits and bounding surfaces in aeolian dune sands. *Sedimentology* 28: 753- 780.

Kocurek, G. & J.Nielson 1986. Conditions favourable for the formation of warm-climate eolian sand sheets. *Sedimentology* 33: 795-816.

Korn, H. & H.Martin 1957. The Pleistocene in South-West Africa. In J.D.Clark (ed.), *Proceedings of 3rd Pan African Congress on Prehistory, Livingstone, 1955:* 14-22.

Lai, R.J. & J.Wu 1978. Wind erosion and deposition along a coastal sand dune. University of Delaware Sea Grant Program, DEL- SG-10-78.

Lancaster, J., N.Lancaster & M.K.Seely 1984. The climate of the central Namib. *Madoqua* 14: 5-61.

Lancaster, N. 1980. The formation of seif dunes from barchans: supporting evidence for Bagnold's model from the Namib Desert. *Z.Geomorph.* 24: 160-167.

Lancaster, N. 1981a. Grain size characteristics of Namib Desert linear dunes. *Sedimentology* 28: 115-122.

Lancaster, N. 1981b. Paleoenvironmental implications of fixed dune systems in southern Africa. *Palaeogeogr.Palaeoclimatol.Palaeoecol.* 33: 327-346.

Lancaster, N. 1982a. Dunes on the Skeleton Coast, Namibia (South West Africa): geomorphology and grain size relationships. *Earth Surface Processes and Landforms* 7: 575-587.

Lancaster, N. 1982b. Linear dunes. *Prog.Phys.Geog.* 6: 476-504.

Lancaster, N. 1982c. Spatial variation in grain size and sorting in the Namib sand sea. *Abstracts of 11th International Congress on Sedimentology. Ontario, Canada*: 62.

Lancaster, N. 1983a. Controls of dune morphology in the Namib sand sea. In M.E.Brookfield & T.S.Ahlbrandt (eds.), *Eolian Sediments and Processes*, Developments in Sedimentology 38. Amsterdam: Elsevier.

Lancaster, N. 1983b. Linear dunes of the Namib sand sea. *Z.Geomorph.Sup.Bd.* 45: 27-49.

Lancaster, N. 1984a. Paleoenvironments in the Tsondab Valley, Central Namib Desert. *Palaeoecology of Africa* 16: 411-419.

Lancaster, N. 1984b. Aridity in southern Africa: age, origins and expression in landforms and sediments. In J.C.Vogel (ed.), *Late Cainozoic Palaeoclimates of the Southern Hemisphere*. Rotterdam: Balkema.

Lancaster, N. 1984c. Aeolian sediments, processes and landforms. *J.Arid Environments.* 7: 249-254.

Lancaster, N. 1985a. Winds and sand movements in the Namib sand sea. *Earth Surface Processes and Landforms* 10: 607-619.

Lancaster, N. 1985b. Variations in wind velocity and sand transport on the windward flanks of desert dunes. *Sedimentology* 32: 581-593.

Lancaster, N. 1986. Grain-size characteristics of linear dunes in the southwestern Kalahari. *J.Sed.Petrol.* 56(3): 395-400.

Lancaster, N. 1987. Variations in wind velocity and sand transport on the windward flanks of desert sand dunes. Reply to comments of A.Watson. *Sedimentology*, 34: 511-520.

Lancaster, N. & C.D.Ollier 1983. Sources of sand for the Namib sand sea. *Z.Geomorph.Sup.Bd.* 45: 71-83.

Lancaster, N., R.Greeley & P.R.Christensen 1987. Dunes of the Gran Desierto sand sea, Sonora, Mexico, *Earth Surface Processes and Landforms*, 12: 277-285.

Lettau, H. & K.Lettau 1969. Bulk transport of sand by the barchan of the Pampa de la Joya in southern Peru. *Z.Geomorph.* 13: 182- 195.

Lindsey, J.F. 1973. Reversing barchan dunes in lower Victoria valley, Antarctica. *Geol.Soc.Amer.Bull.* 84: 1799-1806.

Livingstone, I. 1986a. The dynamics of sand transport on a Namib linear dune. Unpubl. D.Phil Thesis: University of Oxford.

Livingstone, I. 1986b. Geomorphological significance of wind flow patterns over a Namib linear dune. In W.G. Nickling (ed.), *Aeolian Geomorphology*. Allen & Unwin.

Logan, R.F. 1960. *The Central Namib Desert, South West Africa*. Washington DC: National Academy of Sciences.

Lupe, R. & T.S.Ahlbrandt 1979. Sediments of ancient eolian environments: reservoir inhomogenity. In E.D.McKee (ed.), *A Study of Global Sand Seas. US Geol.S.Prof.Paper* 1052: 241-252.

McCoy, F.W., W.J.Nockleberg & R.M.Norris 1967. Speculations on the origin of the Algodones dunes, southern California. *Geol.Soc.Amer.Bull.* 78: 1039-1044.

McKee, E.D. 1966. Structures of dunes at White Sands National Monument, New Mexico (and a comparison with structures of dunes from other selected areas). *Sedimentology* 1: 1-69.

McKee, E.D. 1979. Introduction to a study of global sand seas. In E.D.McKee (ed.), *A Study of Global Sand Seas. US Geol.S.Prof.Paper* 1052: 3-19.

McKee, E.D. 1982. Sedimentary structures in dunes of the Namib Desert, South West Africa. *Geol.Soc.Amer.Spec.Pap.* 186.

McKee, E.D. 1983. Eolian sand bodies of the world. In M.E.Brookfield & T.S.Ahlbrandt (eds.), *Eolian Sediments and Processes*, Developments in Sedimentology 38. Amsterdam: Elsevier.

McKee, E.D. & G.C.Tibbitts 1964. Primary structures of a seif dune and associated deposits in Libya. *J.Sed.Petrol.* 34: 5-17.

McKee, E.D. & C.S.Breed 1976. Sand Seas of the World. US *Geol.S.Prof.Paper* 929: 81-88.

McKee, E.D. & J.J.Bigarella 1979. Ancient sandstones considered to be eolian. In E.D.McKee (ed.), *A Study of Global Sand Seas. US Geol.S.Prof.Paper* 1052: 187-240.

Mabbutt, J.A. 1952. The evolution of the middle Ugab valley, Damaraland, South West Africa. *Trans.Roy.Soc.S.Afr.* 33: 334-366.

Mabbutt, J.A. 1968. Aeolian landforms in central Australia. *Aust.Geog.S.* 6: 139-150.

Mabbutt, J.A. 1977. *Desert Landforms*. Canberra: Australian National Univ.Press.

Mabbutt, J.A. & M.E.Sullivan 1968. Formation of longitudinal dunes, evidence from the Simpson desert. *Aust.Geogr.* 10: 483-487.

Mader, D. 1983. Aeolian sands terminating an evolution of fluvial depositional environments in Middle Bundsandstein (Lower Triassic) of the Eifel, Federal Republic of Germany. In M.E.Brookfield and T.S.Ahlbrandt (eds.), *Eolian Sediments and Processes*, Developments in Sedimentology 38. Amsterdam: Elsevier.

Mader, D. & M.J.Yardley 1985. Migration, modification and merging in aeolian systems and the significance of the depositional mechanisms in Permian and Triassic dune sands of Europe and North America. *Sed.Geol.* 43: 85-218.

Madigan, C.T. 1946. The Simpson Desert Expedition (1939). Scientific Reports 6: Geology. The sand formations. *Trans.R.Soc.S.Aust.* 70: 45-63.

Mainguet, M. 1972. Etude d'un erg (Fachi-Bilma), son alimentation sableuse et sa insertion dans le paysage d'après les photographies prises par satellites. *Académie des Sciences*, C.R. 274: 1633-1636.

Mainguet, M. 1977. Analyse quantitative de l'extremité sahélienne du courant éolien transporteur de sable au Sahara nigérian. *Académie des Sciences, C.R. 285: 1029-1032*.

Mainguet, M. 1978. The influence of trade winds, local air masses and topographic obstacles on the aeolian movement of sand particles and the origin and distribution of ergs in the Sahara and Australia. *Geoforum* 9: 17-28.

Mainguet, M. 1983. Tentative megageomorphological study of the Sahara. In R.Gardner & H.Scoging (eds.), *Megageomorphology*. Oxford: Clarendon Press.

Mainguet, M. 1984. Space observations of Saharan aeolian dynamics. In F.El Baz (ed.), *Deserts and Arid Lands*. The Hague: Nyhoff.

Mainguet, M & L.Canon 1976. Vents et paléovents du Sahara, tentative d'approche paléoclimatique. *Rev.Geogr.Phys.Geol.Dyn.* 18: 241-250.

Mainguet, M. & Y.Callot 1978. L'erg de Fachi-Bilma (Tchad-Niger). *Mémoires et Documents CNRS* 18.

Mainguet, M. & H.C.Chemin 1983. Sand seas of the Sahara and Sahel: an explanation of their thickness and sand dune type by the sand budget principle. In M.E.Brookfield & T.S.Ahlbrandt (eds.), *Eolian Sediments and Processes*, Developments in Sedimentology 38. Amsterdam: Elsevier.

Marker, M.E. 1977. Aspects of the geomorphology of the Kuiseb river, South West Africa. *Madoqua* 10: 199-206.

Marker, M.E. & D.Muller 1978. Relict vlei silts of the middle Kuiseb valley, South West Africa. *Madoqua* 11: 151-162.

Mason, P.J. & R.I.Sykes 1979. Flow over an isolated hill of moderate slope. *Q.J.R.Meteo.S.* 105: 383-395.

Maxwell, T.A. 1982. Sand sheet and lag deposits in the southwestern desert. In F.El-Baz & T.A. Maxwell (eds.), *Desert Landforms of Southwest Egypt: a basis for comparison with Mars*. National Aeronautic and Space Administration Report CR 3611.

Meigs, P. 1966. Geography of coastal deserts. *UNESCO Arid Zones Res.* 28: 1-40.

Merriam, R. 1969. Source of sand dunes of southeastern California and northwestern Sonora, Mexico. *Geol.Soc.Amer.Bull.* 80: 531-534.

Middleton, G.V. & J.B.Southard 1978. Mechanics of sediment transport. *S.E.P.M. Short Course* 3, Dallas, Texas.

Milliman, J.D. & R.H.Meade 1983. World-wide delivery of river sediment to the oceans. *J.Geol.* 91: 1-21.

Moiola, R.J. & D.Weiser 1968. Textural parameters: an evaluation. *J.Sed.Petrol.* 38: 45-53.

Monod, Th. 1958. Majabat Al-Khoubra. *Mem. L'Ins.Franc. D'Af. Noire.*

Nagtegaal, P.J.C. 1973. Adhesion ripples and barchan dune sands of the Recent Namib (South West Africa) and Permian Rotligend (north west Europe) deserts. *Madoqua* 2: 5-19.

Newell, R.E., S.Gould-Stewart & J.C.Chung 1981. Possible interpretation of palaeoclimatic reconstructions for 18 000 BP in the region 60N to 60S, 60W to 100E. *Palaeoecology of Africa* 13: 1-19.

Nielson, J. & G.Kocurek 1986. Climbing zibars of the Algodones. *Sed.Geol.* 48: 1-5.

Norris, R.H. & L.S.Norris 1961. Algodones dunes of southeastern California. *Geol.Soc.Amer.Bull.* 72: 605-620.

Norstrud, H. 1982. Wind flow over a low arbitrary hill. *Boundary Layer Meteorol,* 23:115-124.

Oke, T.R. 1978. *Boundary Layer Climates*. London: Methuen.

Ollier, C.D. 1977. Outline geological and geomorphic history of the central Namib desert. *Madoqua* 10: 207-212.

O'Shea, D.O'C. 1971. An outline of the inshore submarine geology of southern South West Africa and Namaqualand. Unpublished MSc Thesis: University of Cape Town.

Pearse, J.R., D.Lindley & D.C.Stevenson 1981. Wind flow over ridges in simulated atmospheric boundary layers. *Boundary Layer Meteorol.* 21: 77-92.

Porter, M.L. 1986. Sedimentary record of erg migration. *Geology* 14: 497-500.

Rand, R.F. 1920. Angra Pequena (Luderitzbucht) and sub aerial denudation. *Geol.Mag.* 57: 32-35.

Robinson, M.D. & M.K.Seely 1980. Physical and biotic environments of the southern Namib dune ecosystem. *J.Arid Environments 3:* 183- 203.

Rogers, J. 1977. Sedimentation on the continental margins off the Orange River and the Namib Desert. *Joint Geological Survey/University of Cape Town Marine Geoscience Group Bull.* 7.

Rogers, J. 1979. Dispersal of sediment from the Orange River along the Namib desert coast. *S.Afr.J.Sci.* 75: 567.

Rognon, P. 1982. Pluvial and arid phases in the Sahara: the role of non climatic factors. *Palaeoecology of Africa* 12: 45-62.

Rooseboom, A. & H.J.von M.Harmse 1979. Changes in the sediment load of the Orange river during the period 1929-1969. *Int.Assoc.Hydrol., Sci.Publ.* 128: 459-470.

Ross, G.M. 1983. Bigbear Erg: a Proterozoic intermontane eolian sand sea in the Hornby Bay Group, Northwest Territories, Canada. In M.E. Brookfield & T.S.Ahlbrandt (eds.), *Eolian Sediments and Processes*, Developments in Sedimentology 38. Amsterdam: Elsevier.

Royal Navy and South African Air Force 1944. *Weather on the coasts of southern Africa, vol 2, part 1. The West Coast of Africa from River Congo to Olifants River.*

Rubin, D.M. 1984. Factors determining desert dune type. *Nature* 309: 91-92.

Rubin, D.M. & R.E.Hunter 1982. Bedform climbing in theory and nature. *Sedimentology* 29: 121-138.

Rubin, D.M. & R.E.Hunter 1985. Why deposits of longitudinal dunes are rarely recognised in the geologic record. *Sedimentology* 32: 147-157.

Rumpel, D.A. 1985. Successive aeolian saltation: studies of idealised conditions. *Sedimentology* 32: 267-280.

Rust, U. & F.Wienecke 1974. Studies on gramadulla formation in the middle part of the Kuiseb River, South West Africa. *Madoqua* 3: 5-15.

Rust, U. & F. Wienecke 1980. A reinvestigation of some aspects of the evolution of the Kuiseb River valley upstream of Gobabeb, South West Africa. *Madoqua* 12: 163-173.

SACS (South African Committee for Stratigraphy) 1980. Stratigraphy of South Africa: Part 1. *Geological Survey of South Africa, Handbook* 8.

Sarnthein, M. 1978. Sand deserts during glacial maximum and climatic optima. *Nature* 272: 43-46.

Sarnthein, M. & K.Walger 1974. Der äolische Sandstrom aus der w-Sahara zur Atlantikkuste. *Geologische Rundschau* Bd 63: 1065-1087.

Sarnthein, M. & L.Diester Hass 1977. Eolian sand turbidites. *J.Sed.Petrol.* 47: 868-896.

Schenk, C.J. 1983. Textural and structural characteristics of some experimentally formed eolian strata. In M.E.Brookfield & T.S.Ahlbrandt (eds.), *Eolian Sediments and Processes*, Developments in Sedimentology 38. Amsterdam: Elsevier.

Schulze, B.R. 1972. South Africa. In J.F.Griffiths (ed.), *World Survey of Climatology vol 10, Climates of Africa*. Amsterdam: Elsevier.

Schumm, S.A. & R.W.Lichty 1965. Time, space and causality in geomorphology. *A.J.S.* 263: 110-119.

Seely, M.K. 1978. The Namib desert: an unusual ecosystem. *J.Arid Environments* 1: 117-128.

Seely, M.K. & B.H.Sandelowsky 1974. Dating the regression of a river's end point. *S.Afr.Archaeol.B. Goodwin Ser.* 2: 61-64.

Seely, M.K., W.H.Buskirk, W.J.Hamilton & J.E.A.Dixon 1980. Lower Kuiseb River Perennial vegetation survey. *J.South West African Sci.Soc.* 34: 57-86.

Segestrom, K. 1962. Deflated marine terraces as a source of dune chains, Atacama Province, Chile. *US Geol.S.Prof.Paper* 450-C: 91-93.

Selby, M.J. 1977. Bornhardts of the Namib Desert. *Z.Geomorph.* 21: 1-13.

Selby, M.J., C.H.Hendy & M.K.Seely 1979. A late Quaternary lake in the central Namib desert, southern Africa, and some implications. *Palaeogeogr.Palaeoclimatol.Palaeoecol.* 26: 37-41.

Shackley, M. 1980. An Acheulean industry with Elephas reckii fauna from Namib IV, South West Africa (Namibia). *Nature* 284: 340-341.

Sharon, D. 1981. The distribution in space of local rainfall in the Namib desert. *J.Climatol.* 1: 69-75.

Sharp, R.P. 1966. Kelso Dunes, Mohave Desert, California. *Geol.Soc.Amer.Bull.* 77: 1045-1074.

Sharp, R.P. 1980. Wind driven sand in the Coachella Valley, California: further data. *Geol.Soc.Amer.Bull.* 91: 724-730.

Shinn, E.A. 1973. Sedimentary accumulation along the leeward, southeast coast of Qatar Peninsular, Persian Gulf. In B.H. Purser (ed.), *The Persian Gulf*. New York: Springer-Verlag.

Shotton, F.W. 1937. The Lower Bunter sandstones of north Worcestershire and East Shropshire (England). *Geol.Mag.* 74: 534-553

Siesser, W.G. 1978. Aridification of the Namib desert: evidence from oceanic cores. In E.M.Van Zinderen Bakker (ed.), *Antarctic Glacial History and World Palaeoenvironments*. Rotterdam: Balkema.

Siesser, W.G. 1980. Late Miocene origin of the Benguela upwelling system off northern Namibia. *Science* 208: 283-285.

Siesser, W.G. & R.V.Dingle 1981. Tertiary sea level movements around southern Africa. *J.Geol.* 89: 83-96.

Simons, F.S. 1956. A note on Pur-Pur dune, Viru Valley, Peru. *J.Geol.* 64: 517-521.

Smith, H.T.U. 1965. Dune morphology and chronology in central and western Nebraska. *J.Geol.* 73: 557-578.

157

Sneh, A. & T.Weisbrod 1983. Size-frequency distribution on longitudinal dune ripple flank sands compared to that of slipface sands of various dune types. *Sedimentology* 30: 717-726.

South African Weather Bureau 1975. *Climate of South Africa part 12, Surface Winds.*

Steele, R.P. 1983. Longitudinal draa in the Permian Yellow sands of north-east England. In M.E.Brookfield & T.S.Ahlbrandt (eds.), *Eolian Sediments and Processes*, Developments in Sedimentology 38. Amsterdam: Elsevier.

Stengel, H.W. 1964. The rivers of the Namib and their discharge into the Atlantic, part I: Kuiseb and Swakop. *Scientific Papers of the Namib Desert Research Station* 22: 1-50.

Stengel, H.W. 1970. Die riviere van die Namib met hulle toelope na die Atlantiese Oseaan. *Derde Deel: Tsondab, Tsams en Tsauchab*. Unpublished Report: Department of Water Affairs, South West Africa Branch.

Stokes, W.L. 1964. Eolian varving in the Colorado Plateau. *J.Sed.Petrol.* 34: 429-432.

Street, F.A. 1981. Tropical palaeoenvironments. *Prog.Phys.Geog.* 5: 157-185.

Striem, H.L. 1954. The seifs on the Israel-Sinai border and the correlation of their alignment. *Research Council of Israel Bull.*

Sweeting, M.M. & N.Lancaster 1982. Solutional and wind erosion forms on limestone in the central Namib desert. *Z.Geomorph.* 26: 197-207.

Talbot, M.R. 1980. Environmental responses to climatic change in the west African Sahel over the past 20 000 years. In M.A.J.Williams & H.Faure (eds.), *The Sahara and the Nile*. Rotterdam: Balkema.

Talbot, M.R. 1985. Major bounding surfaces in aeolian sandstone: a climatic model. *Sedimentology* 32: 257-266.

Tankard, A.J. & J.Rogers 1978. Late Cenozoic palaeoenvironments on the west coast of southern Africa. *J.Biogeog.* 5: 319-337.

Tankard, A.J., M.P.A.Jackson, K.A.Eriksson, D.K.Hobday, D.R.Hunter & W.E.L.Minter 1982. *Crustal Evolution of Southern Africa: 3.8 Billion Years of Earth History*. New York: Springer Verlag.

Tanner, W.F. 1965. Upper Jurassic paleogeography of the Four Corners Region. *J.Sed.Petrol.* 35: 564-574.

Teller, J.T. & N.Lancaster 1985. History of sediments at Khommabes, central Namib desert. *Madoqua* 14(3): 267-278.

Teller, J.T. & N. Lancaster 1986a. Interdune lacustrine deposits of the Namib Sand Sea, Namibia. Abstracts *12th International Congress on Sedimentology, Canberra, Australia : 98.*

Teller J.T. & N.Lancaster 1986b. Lacustrine sediments at Narabeb in the central Namib desert, Namibia. *Paleogeogr.Paleoclimatol.Paleoecol.* 56: 177-195.

Torquato, J.R. 1972. Origin and evolution of the Mocamedes desert (Angola). In *African Geology: Quaternary Rocks and Geomorphology of Angola, Chad, Cote d'Ivoire, Nigeria and Sahara*. Ibadan: Department of Geography, University of Ibadan.

Tseo, G. 1986. Longitudinal dunes: their genesis and ordering. Unpubl. PhD Thesis: University of Adelaide.

Tsoar, H. 1974. Desert dune morphology and dynamics El Arish (northern Sinai). *Z.Geomorph.Sup.Bd.* 20: 41-61.

Tsoar, H. 1978. *The Dynamics of Longitudinal Dunes*. Final Technical Report: U.S. Army European Research Office.

Tsoar, H. 1982. Internal structure and surface geometry of longitudinal (seif) dunes. *J.Sed.Petrol.* 52: 823-831.

Tsoar, H. 1983. Dynamic processes acting on a longitudinal (seif) dune. *Sedimentology* 30: 567-578.

Tsoar, H. 1985. Profiles analysis of sand dunes and their steady state significance. *Geogr.Ann.* 67A: 47-59.

Tsoar, H., R.Greeley & A.R.Peterfreund 1979. Mars: the north polar sand sea and related wind patterns. *J.G.R.* 84: 8167-8180.

Twidale, C.R. 1972. Evolution of sand dunes in the Simpson Desert, central Australia. *Trans.Inst.Brit.Geog.* 56: 77-110.

Tyson, P.D. & M.K.Seely 1980. Local winds over the central Namib. *S.Afr.Geog.J.* 62: 136-150.

Udden, J.A. 1898. *Mechanical composition of wind deposits.* Augustana Library Publication.

Van Zinderen Bakker, E.M. 1975. The origin and paleoenvironment of the Namib desert biome. *J.Biogeog.* 2: 65-74.

Van Zinderen Bakker, E.M. 1984a. Aridity along the Namibian coast. *Palaeoecology of Africa* 16: 149-162.

Van Zinderen Bakker, E.M. 1984b. A late and post-glacial pollen record from the Namib desert. *Palaeoecology of Africa* 16: 421- 428.

Vincent, P.J. 1984. Particle size variation over a transverse dune in the Nafid as Sirr, central Saudi Arabia. *J.Arid Environments* 7: 329-336.

Visher, G.S. 1969. Grain size distributions and depositional processes. *J.Sed.Petrol.* 39: 1074-1106.

Vogel, J.C. 1982. The age of the Kuiseb River silt terrace at Homeb. *Palaeoecology of Africa* 15: 201-209.

Vogel, J.C. & E.Visser 1981. Pretoria radiocarbon dates II. *Radiocarbon* 23: 43-80.

Walker, R.G. & G.V.Middleton 1981. Facies models 4; Eolian sands. In R.G. Walker (ed.), *Facies Models.* Geosci.Can. 1.

Walker, T.R. 1979. Red color in dune sand. In E.D.McKee (ed.), *A Study of Global Sand Seas. US Geol.S.Prof.Paper* 1052: 61-82.

Walmsley, J.L. & A.D.Howard 1985. Application of a boundary layer model to flow over an eolian dune. *J.G.R.* 90: 10631-10640.

Ward, J.D. 1982. Aspects of a suite of Quaternary conglomeratic sediments in the Kuiseb valley, Namibia. *Palaeoecology of Africa* 15: 211-216.

Ward, J.D. 1984. Aspects of the Cenozoic Geology in the Kuiseb Valley, central Namib Desert. Unpubl. PhD Thesis: University of Natal.

Ward, J.D. & V.Von Brunn 1985. Sand dynamics along the Kuiseb River. In B.J.Huntley (ed.), *The Kuiseb Environment: the development of a monitoring baseline.* Pretoria: South African National Scientific Programmes Report 106.

Ward, J.D., M.K.Seely & N.Lancaster 1983. On the antiquity of the Namib. *S.Afr.J.Sci.* 79: 175-183.

Warren, A. 1970. Dune trends and their implications in central Sudan. *Z.Geomorph.Sup.Bd.* 10: 154-180.

Warren, A. 1972. Observations on dunes and bimodal sands in the Tenere desert. *Sedimentology* 19: 37-44.

Warren, A. 1976. Dune trend and the Ekman Spiral. *Nature* 259: 653-654.

Warren, A. 1979. Aeolian processes. In C.Embleton & J.Thornes (eds.), *Processes in Geomorphology.* London: Edward Arnold.

Warren, A. 1984. Progress report on arid geomorphology. *Prog.Phys.Geog.* 8: 399-420.

Warren, A. & P.Knott 1983. Desert dunes: a short review of needs in desert dune research and a recent study of micro-meteorological dune initiation mechanisms. In M.E.Brookfield & T.S.Ahlbrandt (eds.), *Eolian Sediments and Processes*, Developments in Sedimentology 38. Amsterdam: Elsevier.

Wasson, R.J. 1983a. The Cainozoic history of the Strzelecki and Simpson dunefields (Australia) and the origin of desert dunes. *Z.Geomorph.Sup.Bd.* 45: 85-115.

Wasson, R.J. 1983b. Dune sediment type, sand colour, sediment provenence and hydrology in the Strzelecki-Simpson dunefield, Australia. In M.E.Brookfield & T.S.Ahlbrandt (eds.), *Eolian Sediments and Processes*, Developments in Sedimentology, 38. Amsterdam: Elsevier.

Wasson, R.J. & R.Hyde 1983a. Factors determining dune types. *Nature* 304: 337-339.

Wasson, R.J. & R.Hyde 1983b. A test of granulometric control of desert dune geometry. *Earth Surface Processes and Landforms* 8: 301-312.

Wasson, R.J., S.N.Rajaguru, V.N.Misra, D.P.Agrawal, R.P.Dhir, A.K.Srigrivi & K.Kameswara 1983. Geomorphology, late Quaternary stratigraphy and paleoclimatology of the Thar dunefield. *Z.Geomorph.Sup.Bd.* 45: 117-151.

Watson, A. 1979. Gypsum crusts in deserts. *J.Arid Environments* 2: 3-20.

Watson, A. 1986. Grain size variations on a longitudinal dune and a barchan dune. *Sed.Geol.* 46: 49-66.

Wellington, J.H. 1955. *Southern Africa: A Geographical Study, vol 1, Physical Geography.* Cambridge: Cambridge Univ.Press.

Whitney, J.W., D.J.Faulkender & H.Rubin 1983. The Environmental History and Present Condition of the Northern Sand Seas of Saudi Arabia. Open file report United States Geological Survey. Ministry of Petroleum and Mineral Resources Deputy Minister for Mineral Resources, Jeddah, Saudi Arabia.

Wilmer, H.C. 1883. The relationship of the sand dune formation on the south west coast of Africa to local wind currents. *Trans.S.Afr.Phil.Soc.* 5: 326-329.

Wilson, I.G. 1971. Desert sandflow basins and a model for the development of ergs. *Geogr.J.* 137: 180-197.

Wilson I.G. 1972. Aeolian bedforms, their development and origins. *Sedimentology* 19: 173-210.

Wilson, I.G. 1973. Ergs. *Sed.Geol.* 10: 77-106.

Wolman, M.G. & J.P.Miller 1960. Magnitude and frequency of forces in geomorphic processes. *J.Geol.* 68: 54-74.

Wopfner, A. & G.R.Twidale 1967. Geomorphological history of the Lake Eyre basin. In J.N.Jennings & J.A.Mabbutt (eds.), *Landform Studies from Australia and New Guinea.* Cambridge: Cambridge Univ.Press.

Yaalon, D.H. & J.D.Ward 1982. Observations on calcrete and recent calcic horizons in relation to landforms in the central Namib desert. *Palaeoecology of Africa* 15: 183-186.

Yalin, M.S. 1972. *Mechanics of Sediment Transport.* Oxford: Pergamon Press.

APPENDICES

Appendix 1: Heavy mineral composition of dune sands.
Summary table of heavy minerals (major = approx > 20%; minor = < 20%; rare = a few grains) each column is in approximate order of abundance.

Sample	Major	Minor	Rare
I	garnet (large)	opaques clinopyroxene	
II	garnet (large) clinopyroxene opaques		
III	clinopyroxene opaques	garnet	hornblende zircon biotite
IV	clinopyroxene opaques	garnet hornblende	tourmaline zircon rutile
V	clinopyroxene opaques garnet (large)	hornblende	zircon
VI	clinopyroxene opaques	garnet hornblende	zircon
VII	clinopyroxene opaques	garnet	apatite
VIII	opaques (large) garnet (large)	clinopyroxene (small) opaques (small)	epidote hornblende chlorite zircon
IX	clinopyroxene opaques	garnet	zircon hornblende epidote
X (north)	clinopyroxene opaques	garnet	rutile hornblende
X (south)	garnet (large) opaques		epidote
XI	clinopyroxene opaques	garnet	epidote staurolite hornblende tourmaline apatite
XII	clinopyroxene opaques	garnet (large)	hornblende epidote zircon
XII a	garnet (large)	opaques (large)	clinopyroxene hornblende staurolite
XIII	clinopyroxene opaques	garnet	apatite hornblende epidote
XIV	garnet (large)	opaques clinopyroxene	hornblende zircon

162

Appendix 1 (continued)

Sample	Major	Minor	Rare
XV	clinopyroxene	garnet opaques	hornblende rutile
XVI	garnet (large)	clinopyroxene (very small) opaques (very small) hornblende	hypersthene
XVII	clinopyroxene opaques	garnet staurolite	hornblende epidote tourmaline
XVIII	clinopyroxene garnet opaques	staurolite	tourmaline epidote zircon
XIX	garnet (large) opaques clinopyroxene	hornblende	epidote biotite rutile staurolite chlorine zircon tourmaline
XXI	clinopyroxene garnet (large) opaques	hornblende	biotite epidote staurolite rutile
XXII	clinopyroxene garnet opaques		zircon staurolite hornblende
XXIII	clinopyroxene opaques hornblende	garnet (large)	biotite staurolite tourmaline
XXIV	clinopyroxene garnet opaques	hornblende	biotite epidote rutile
XXV	clinopyroxene garnet opaques	hornblende staurolite	tourmaline
XXVI	clinopyroxene garnet opaques	hornblende	epidote hypersthene

163

Appendix 2: Mean values of grain size and sorting parameters for each site (phi units). Parameters calculated using graphical measures of Folk and Ward (1957). For location of sites see Fig. 21.

Site	Mean	Standard deviation	Skewness	Kurtosis
Site 1				
Linear				
crest	2.55	0.45	0.05	0.50
mid slip	2.42	0.38	0.04	0.54
base slip	2.18	0.55	0.08	0.50
E plinth	2.21	0.66	0.01	0.47
W plinth	2.01	0.68	0.32	0.45
u west	2.17	0.62	0.18	0.49
interdune	1.57	1.02	0.39	0.45
Crossing				
crest	2.11	0.69	0.28	0.51
Zibar	1.89	0.93	0.12	0.51
Site 1A				
Linear				
crest	2.48	0.43	0.14	0.55
plinth	2.14	0.48	0.06	0.48
Barchan				
crest	2.35	0.48	0.15	0.52
mid slip	2.28	0.38	−0.03	0.54
base slip	2.19	0.47	−0.06	0.60
base stoss	2.36	0.37	0.03	0.57
horn	2.36	0.47	−0.10	0.54
Site 2				
Linear				
crest	2.33	0.36	0.25	0.57
mid slip	2.34	0.40	0.00	0.54
base slip	2.27	0.48	0.04	0.50
E plinth	2.01	0.56	0.34	0.55
W plinth	2.12	0.64	0.45	0.49
u west	2.23	0.60	0.32	0.51
interdune	1.98	0.71	0.26	0.54
Crossing				
crest	2.45	0.39	0.07	0.52
mid slip	2.43	0.38	−0.03	0.54
Site 3				
Linear				
crest	2.40	0.46	0.18	0.53
mid slip	2.53	0.44	0.08	0.52
base slip	2.14	0.54	0.21	0.49
E plinth	2.29	0.60	−0.19	0.53
W plinth	2.09	0.84	0.50	0.42
u west	2.42	0..60	−0.04	0.50
interdune	2.19	0.75	−0.05	0.45

Site	Mean	Standard deviation	Skewness	Kurtosis
Crossing				
crest	2.05	0.59	0.14	0.47
mid slip	2.45	0.47	−0.05	0.52
base slip	2.00	0.68	0.11	0.45
u west	2.08	0.64	0.01	0.47
Zibar	2.15	0.99	0.14	0.40
Site 4				
Linear				
crest	2.41	0.30	0.14	0.51
mid slip	2.52	0.30	0.03	0.49
base slip	2.27	0.36	0.09	0.52
E plinth	2.00	0.54	0.42	0.67
W plinth	2.16	0.94	0.27	0.39
u west	2.25	0.65	0.32	0.49
interdune	3.11	1.13	−0.16	0.59
Crossing				
crest	2.16	0.33	0.10	0.53
mid slip	2.27	0.29	0.09	0.55
base slip	1.97	0.41	0.44	0.56
Site 5				
Linear				
crest	2.68	0.40	0.12	0.54
mid slip	2.53	0.29	0.04	0.48
base slip	2.32	0.26	−0.04	0.51
E plinth	2.34	0.58	0.25	0.55
W plinth	2.21	0.79	0.29	0.46
u west	2.31	0.67	−0.03	0.45
u east	2.22	0.48	0.08	0.50
interdune	2.28	0.24	0.26	0.53
Site 6				
Linear				
crest	2.40	0.40	0.17	0.50
mid slip	2.56	0.40	0.07	0.50
base slip	2.30	0.45	0.17	0.51
E plinth	2.10	0.91	0.25	0.44
W plinth	2.20	0.91	0.42	0.42
u west	2.43	0.52	0.17	0.48
u west	2.41	0.68	0.04	0.47
interdune	2.13	0.84	0.30	0.44
E flank				
barchanoid	2.40	0.47	0.09	0.54
Crossing				
crest	2.21	0.45	0.32	0.54
mid slip	2.55	0.52	0.21	0.45
Zibar	2.17	0.97	0.45	0.42

Appendix 2 (continued)

Site	Mean	Standard deviation	Skewness	Kurtosis
Site 7				
Linear				
crest	2.56	0.38	0.09	0.50
mid slip	2.59	0.37	−0.02	0.50
base slip	2.19	0.50	0.19	0.47
E plinth	2.10	0.78	0.35	0.46
W plinth	2.15	1.01	0.27	0.42
u west	2.35	0.59	0.08	0.47
u east	2.48	0.46	0.26	−/49
interdune	1.78	1.07	0.50	0.44
E flank				
barchanoid	2.50	0.48	0.24	0.51
Crossing				
crest	2.40	0.49	0.16	0.53
Zibar	2.23	0.97	0.22	0.41
Site 8				
Linear				
crest	2.45	0.34	0.13	0.51
mid slip	2.60	0.33	0.04	0.52
base slip	2.32	0.37	0.08	0.52
E plinth	2.06	0.63	0.35	0.50
W plinth	1.86	0.77	0.57	0.45
u west	2.31	0.48	0.17	0.50
u east	2.34	0.54	0.18	0.55
interdune	2.03	1.03	0.42	0.43
E flank				
barchanoid	2.31	0.32	0.15	0.52
Crossing				
crest	1.98	0.27	0.32	0.61
Zibar	1.88	0.72	0.40	0.45
Site 9				
Linear				
crest	2.36	0.43	0.22	0.54
mid slip	2.44	0.54	−0.06	0.53
E plinth	2.24	0.57	0.32	0.44
W plinth	2.10	0.98	0.46	0.43
u west	2.19	0.79	0.27	0.45
interdune	2.01	1.09	0.26	0.46
E flank				
barchanoid	2.23	0.62	0.18	0.48
Crescentic				
crest	2.07	0.51	0.30	0.54
mid slip	2.39	0.55	0.06	0.50
base slip	2.02	0.58	0.18	0.49
base stoss	2.18	1.04	0.26	0.39

Site	Mean	Standard deviation	Skewness	Kurtosis
mid stoss	2.22	0.71	0.19	0.47
Crossing				
crest	2.50	0.26	0.22	0.62
Zibar	2.29	0.86	0.38	0.45
Site 10				
Star				
crest	2.23	0.28	0.16	0.55
mid slip	2.46	0.32	0.20	0.54
base slip	2.52	0.34	0.06	0.54
E plinth	2.25	0.59	0.19	0.53
W plinth	2.19	0.50	0.16	0.55
u west	2.32	0.32	0.20	0.56
u east	2.36	0.38	0.28	0.55
interdune	2.05	0.83	0.04	0.52
E flank				
barchanoid	2.37	0.21	0.23	0.52
W flank				
barchanoid	2.25	0.31	0.14	0.55
Barchan				
crest	2.27	0.28	0.19	0.59
Crossing				
crest	2.16	0.30	0.35	0.63
Zibar	1.62	1.15		0.40
Site 11				
Crescentic				
crest	2.51	0.50	0.02	0.48
mid slip	2.56	0.47	−0.01	0.50
base slip	2.53	0.47	−0.05	0.49
u east	2.29	0.57	0.20	0.46
base stoss	2.26	0.59	0.10	0.49
mid stoss	2.42	0.65	−0.02	0.48
Barchan				
crest	2.15	0.67	−0.14	0.41
Site 12				
Linear				
crest	2.45	0.43	0.07	0.47
mid slip	2.57	0.43	0.02	0.46
E plinth	1.97	0.74	0.27	0.46
W plinth	2.15	0.90	0.36	0.42
u west	2.45	0.58	0.11	0.49
u east	2.36	0.42	0.01	0.51
interdune	1.73	0.87	0.46	0.52
E flank				
barchanoid	2.32	0.51	0.22	0.50

Appendix 2 (continued)

	Mean	Standard deviation	Skewness	Kurtosis
Crossing				
crest	2.27	0.51	0.24	0.52
Site 12A				
Linear				
crest	2.01	0.35	0.24	0.57
mid slip	2.37	0.45	−0.01	0.48
E plinth	2.11	0.89	0.06	0.48
W plinth	1.83	0.71	0.35	0.52
u west	2.55	0.54	0.04	0.52
u east	2.18	0.54	0.09	0.47
interdune	2.21	0.99	0.56	0.48
Site 13				
Linear				
crest	2.50	0.32	0.04	0.46
mid slip	2.47	0.35	−0.04	0.47
E plinth	1.98	0.94	0.46	0.45
W plinth	1.98	0.92	0.59	0.51
u west	2.49	0.32	0.04	0.51
u east	2.60	0.38	0.16	0.50
interdune	1.74	1.01	0.44	0.47
E flank				
barchanoid	2.16	0.41	0.20	0.51
Site 14				
Star				
crest	2.21	0.28	0.07	0.47
mid slip	2.35	0.29	0.05	0.47
E plinth	1.91	0.47	0.06	0.68
u west	2.40	0.34	0.03	0.49
u east	2.16	0.36	0.19	0.49
interdune	1.84			
E flank				
barchanoid	1.83	0.30	0.16	0.53
Crossing				
crest	2.18	0.35	0.17	0.51
Site 15				
Linear				
crest	2.06	0.48	0.23	0.51
mid slip	2.39	0.35	−0.01	0.48
E plinth	1.86	0.88	0.44	0.46
W plinth	2.05	0.90	0.17	0.43
u west	2.54	0.50	−0.04	0.61
interdune	1.95	0.91	0.26	0.46
Crossing				
crest	2.13	0.54	−0.01	0.45

Appendix 2 (continued)

	Mean	Standard deviation	Skewness	Kurtosis
Site 16				
Zibar				
crest	1.99	0.93	0.44	0.46
interdune	1.83	0.89	0.43	0.48
mid stoss	1.84	0.87	0.34	0.45
Site 16A				
Linear				
crest	2.22	0.37	0.24	0.51
E plinth	1.97	1.10	0.14	0.47
W plinth	2.01	1.25	0.27	0.40
interdune	2.15	1.17	0.13	0.40
Site 17				
Linear				
crest	2.40	0.36	0.13	0.49
mid slip	2.50	0.38	0.01	0.45
E plinth	1.90	1.00	0.28	0.44
W plinth	2.11	0.73	−0.02	0.49
interdune	1.93	0.96	0.09	0.46
E flank				
barchanoid	2.30	0.35	0.24	0.54
Site 18				
Linear				
crest	2.30	0.47	0.25	0.51
mid slip	2.24	0.37	0.02	0.49
e plinth	1.98	0.63	0.26	0.51
W plinth	2.11	0.63	0.27	0.54
u west	2.05	0.55	0.36	0.49
interdune	2.25	0.78	0.14	0.50
E flank				
barchanoid	2.27	0.22	0.09	0.53
Site 19				
Star				
crest	2.57	0.25	0.07	0.51
mid slip	2.55	0.25	0.03	0.50
base slip	2.40	0.24	0.09	0.50
E plinth	2.33	0.48	0.18	0.56
W plinth	2.33	0.52	0.21	0.52
u west	2.54	0.32	0.06	0.50
u east	2.44	0.36	0.12	0.53
interdune	2.20	0.81	0.18	0.47
E flank				
barchanoid	2.46	0.38	0.17	0.52
Site 20				
Crescentic				
crest	2.10	0.41	0.30	0.54

Appendix 2 (continued)

	Mean	Standard deviation	Skewness	Kurtosis
mid stoss	2.57	0.65	0.26	0.51
Barchan				
crest	2.22	0.48	0.25	0.51
Sand sheet	1.84	1.25	0.49	0.38
Zibar	2.07	1.12	0.51	0.39
Site 21				
Crescentic				
crest	2.04	0.78	0.30	0.46
mid slip	2.45	0.67	−0.22	0.47
mid stoss	2.09	0.83	0.28	0.45
Zibar	1.96	1.08	0.30	0.42
Site 21A				
Crescentic				
crest	2.01	0.70	0.32	0.52
Site 22A				
Crescentic				
crest	2.23	0.69	0.16	0.43
mid stoss	2.10	0.93	0.30	0.39
Site 23				
Crescentic				
crest	2.25	0.45	0.07	0.48
Barchan				
crest	2.25	0.45	0.07	0.48
mid slip	2.55	0.32	−0.03	0.51
mid stoss	2.29	0.55	−0.33	0.52
Site 24				
Star				
crest	2.25	0.34	0.10	0.53
W plinth	1.96	0.65	0.11	0.46
u west	1.94	0.44	0.20	0.48
u east	2.11	0.37	0.09	0.50
star arm	2.17	0.36	0.16	0.51
Sand sheet	2.02	1.03	0.22	0.45
Site 25				
Linear				
crest	2.16	0.51	0.32	0.52
E plinth	2.02	0.95	0.30	0.45
W plinth	2.03	0.91	0.17	0.44
interdune	1.98	0.89	0.26	0.45
Crossing				
crest	2.43	0.45	0.14	0.34
Site 26				
Linear				
crest	2.14	0.37	0.27	0.49

Appendix 2. (continued)

	Mean	Standard deviation	Skewness	Kurtosis
E plinth	2.11	0.53	0.27	0.49
W plinth	2.04	0.67	0.31	0.49
u west	1.95	0.74	0.33	0.47
interdune	1.97	0.70	0.33	0.48
E flank barchanoid	2.06	0.39	0.28	0.52

Appendix 3: Mean values of grain size frequency for each site, by dune type and position.

% in each size class (phi units)									
	0.0	0.5	1.0	1.5	2.0	2.5	3.0	3.5	4.0
Site 1									
Linear									
crest				1.13	11.47	35.93	39.74	10.25	1.61
mid slip				2.03	11.47	43.44	35.15	7.06	0.74
base slip				11.87	27.62	29.67	21.09	7.69	1.43
E plinth				15.31	21.85	26.97	23.56	9.60	2.17
W plinth			1.27	28.66	27.85	17.79	12.24	8.23	3.04
u west				17.86	20.65	19.56	17.75	6.31	2.68
interdune		22.31	11.24	17.74	13.61	12.67	9.42	7.11	3.81
Crossing									
crest			0.21	17.73	34.06	21.68	12.99	8.47	3.73
Zibar		7.98	6.64	16.69	27.20	19.82	7.69	6.04	5.09
Site 1A									
Linear									
crest				1.06	8.44	45.16	33.75	8.70	2.38
plinth				6.11	33.35	37.74	17.97	4.16	0.58
Barchan									
crest				1.77	19.90	43.12	24.14	8.18	2.61
mid slip			0.13	3.77	15.30	54.70	22.63	2.80	0.54
base slip			0.78	8.45	18.04	57.32	16.46	3.47	1.15
base stoss				0.92	13.47	53.32	27.75	3.87	0.56
horn				4.29	19.09	37.11	30.37	7.86	1.02
Site 2									
Linear									
crest					19.28	41.48	29.54	5.82	0.95
mid slip					16.38	38.83	93.08	5.76	0.66
base slip				0.75	18.60	35.18	29.62	5.78	0.88
E plinth				15.82	38.80	27.13	10.21	4.90	2.13
W plinth				37.02	18.45	20.15	18.45	7.78	4.21
u west			2.11	12.29	26.29	27.69	20.37	8.93	3.27
interdune		2.00	5.60	20.34	29.86	9.63	6.28	4.06	2.29
Crossing									
crest				0.57	10.89	45.74	33.04	7.91	1.60
mid slip				1.55	12.25	41.62	37.10	6.61	0.77
Site 3									
Linear									
crest			0.23	4.14	16.73	40.96	26.16	10.21	2.08
mid slip				2.22	9.35	35.11	37.05	14.16	1.71
base slip			1.89	12.78	27.75	32.69	16.99	7.21	1.01
E plinth			2.96	9.77	72.45	37.94	25.64	9.31	1.44
W plinth			1.79	36.97	16.36	11.27	13.58	12.55	4.62
u west			1.86	7.81	13.43	30.84	28.62	14.31	2.71
interdune		0.58	4.42	22.73	14.53	22.14	21.68	11.04	2.64
Crossing									
crest			0.38	12.83	22.72	18.52	6.51	2.13	1.79

Appendix 3 (continued)

% in each size class (phi units)								
0.0	0.5	1.0	1.5	2.0	2.5	3.0	3.5	4.0
mid slip		0.09	3.70	12.12	38.73	33.88	10.26	1.09
base slip		3.11	25.37	22.81	25.19	14.36	7.10	1.44
Zibar	0.15	10.87	22.90	14.79	13.95	11.73	16.92	7.18
Site 4								
Linear								
crest			0.84	10.65	56.27	29.20	3.18	0.25
mid slip			0.72	4.59	41.69	48.27	4.70	0.33
base slip			1.46	22.46	51.61	20.98	3.07	0.55
E plinth			9.82	50.53	23.55	6.56	5.83	2.61
W plinth		1.36	37.84	10.48	9.52	16.58	13.35	6.77
u west		0.21	10.66	32.49	23.90	17.09	9.69	4.47
interdune	0.38	3.51	6.73	5.97	17.33	23.06	18.58	
Crossing								
crest			6.19	32.62	37.96	19.17	3.50	0.45
mid slip			2.96	23.48	45.77	24.38	3.92	0.75
base slip			18.73	38.88	30.67	7.44	3.16	0.78
Site 5								
Linear								
crest			0.34	4.59	26.47	47.93	17.15	2.92
mid slip				4.86	40.50	47.64	6.46	0.33
base slip			0.35	22.56	47.12	28.44	1.45	0.06
E plinth			8.04	23.77	30.36	22.54	10.14	3.76
W plinth	0.09	1.22	17.15	30.02	19.08	13.58	11.83	5.30
u west			15.08	18.45	21.50	27.21	13.71	3.01
u east			5.66	30.13	48.60	19.43	4.95	0.94
interdune	2.16	1.41	7.12	28.15	28.39	14.88	9.63	5.18
Site 6								
Linear								
crest			2.51	17.07	41.24	28.93	9.55	1.26
mid slip			1.85	8.39	33.29	41.41	14.45	1.25
base slip		0.62	6.33	20.88	36.35	24.13	7.43	0.70
E plinth		7.48	22.66	19.14	16.76	12.66	13.17	5.29
W plinth		2.12	28.85	21.06	10.72	10.50	15.10	7.45
u west		0.98	8.31	15.74	30.85	27.87	14.03	2.98
u east		2.63	10.23	16.07	22.87	26.16	17.01	3.98
interdune	0.54	3.94	23.08	24.09	16.41	12.02	13.16	5.19
E flank								
barchanoid		0.26	2.50	13.86	43.85	30.31	8.34	1.21
Crossing								
crest		0.87	7.39	24.68	41.13	17.13	7.36	1.64
mid slip			0.46	12.80	38.69	34.51	12.51	0.97
Zibar	0.59	5.85	21.93	26.22	17.86	11.94	10.17	3.69
Site 7								
Linear								
crest			3.21	8.64	29.36	45.06	13.99	1.34

Appendix 3 (continued)

| % in each size class (phi units) | | | | | | | | |
0.0	0.5	1.0	1.5	2.0	2.5	3.0	3.5	4.0
mid slip			0.59	6.73	31.70	47.37	12.75	0.90
base slip		1.59	10.77	29.28	29.84	22.93	6.94	0.76
E plinth		4.30	21.72	27.28	17.15	13.24	10.83	3.78
W plinth		4.74	21.76	12.18	8.54	12.17	11.09	7.04
u west		0.65	10.04	21.05	26.55	27.93	12.26	2.26
u east		0.97	17.00	14.13	16.33	35.70	16.60	3.37
interdune	5.80	25.19	23.88	10.93	6.05	8.37	8.90	6.60
E flank								
barchanoid			2.31	16.30	33.54	24.99	11.62	3.44
Crossing		1.87	7.94	12.50	38.52	28.46	10.98	2.01
Zibar		7.35	21.56	18.54	12.45	12.03	16.31	8.47
Site 8								
Linear								
crest			0.70	14.01	37.33	37.94	8.58	0.70
mid slip			0.27	5.89	30.77	50.88	11.42	0.78
base slip			0.96	19.43	47.82	26.89	4.35	0.55
E plinth		1.35	28.88	34.80	15.72	9.16	6.35	2.55
W plinth		1.96	46.21	18.64	7.20	9.08	9.59	4.54
u west			7.23	18.75	27.53	33.26	11.75	1.87
u east			5.66	20.58	40.70	19.53	6.62	2.96
interdune	0.74	11.05	30.31	16.72	10.30	8.73	9.68	6.87
E flank								
barchanoid			1,12	23,20	43,40	25,00	6.11	1.08
Crossing								
crest			2.13	59.17	31.59	4.84	1.79	0.41
Zibar		4.83	35.88	21.10	8.61	8.15	11.13	6.97
Site 9								
Linear								
crest		0.27	5.99	19.97	37.33	26.08	10.42	1.62
mid slip			3.90	16.20	34.51	30.66	12.72	1.68
E plinth		2.40	20.22	12.32	25.28	25.91	12.15	2.44
W plinth		5.43	31.79	20.29	9.60	9.33	11.57	7.16
u west		2.14	22.07	24.18	16.33	16.06	12.08	4.70
interdune	6.66	4.31	8.81	23.60	16.84	9.86	11.06	12.60	7.64
E flank								
barchanoid		0.24	9.07	31.87	27.78	18.30	9.26	2.54
Star								
crest				0.53	35.21	45.06	16.66	2.33
Crescentic								
crest		1.76	18.10	30.61	23.97	14.62	8.30	2.19
mid slip		0.42	6.12	19.99	31.48	26.44	12.96	2.47
base slip		4.94	17.37	28.71	25.77	15.30	6.84	1.43
base stoss		9.56	27.38	13.50	7.93	10.49	17.17	9.25
mid stoss			2.25	15.86	23.81	17.91	11.98	3.60
Crossing								

Appendix 3 (continued)

% in each size class (phi units)								
0.0	0.5	1.0	1.5	2.0	2.5	3.0	3.5	4.0
crest			0.04	3.60	57.10	32.36	6.02	0.81
Zibar	0.08	1.76	18.90	26.17	16.97	13.40	14.82	6.26

Site 10
Star

crest			5.23	24.70	51.76	19.16	2.53	0.38
mid slip			2.41	7.20	51.78	33.13	6.48	0.84
base slip			0.46	10.29	41.17	36.63	9.67	1.06
E plinth		0.81	13.89	22.92	33.44	16.64	6.76	2.36
W plinth		1.78	9.09	27.51	33.90	21.00	5.89	1.39
u west			7.13	16.91	52.06	24.97	3.98	0.72
u east			0.39	11.71	44.94	25.99	5.71	1.55
interdune	4.55	4.63	5.19	11.61	21.31	22.15	20.01	8.26	3.64

E flank

barchanoid			0.02	3.46	72.87	22.13	1.28	0.17

Barchan

crest		0.12	1.56	23.95	47.29	20.81	5.62	0.78

Site 11
Crescentic

crest		0.24	4.54	15.17	19.51	42.09	16.31	2.38
mid slip			3.72	12.21	16.88	46.14	18.69	2.16
base slip			3.13	11.15	19.29	49.17	15.84	1.28
mid stoss		0.36	15.51	15.80	12.31	34.11	17.93	3.49
base stoss		0.32	18.51	17.17	17.69	33.86	10.83	1.42

Barchan

crest		1.61	2.84	18.66	14.61	25.82	14.89	2.45

Site 12
Linear

crest			2.04	15.71	30.08	40.25	10.57	2.12
mid slip			0.42	9.95	26.85	46.56	12.95	2.75
E plinth	1.83	6.42	22.64	26.16	14.85	16.53	8.16	3.19
W plinth		3.55	29.23	20.40	8.26	14.81	12.16	8.16
u west		0.32	5.73	16.92	23.08	36.27	12.45	4.45
u east			6.49	19.28	30.94	41.32	6.61	1.05
interdune	2.74	16.07	33.82	16.80	6.85	9.76	7.32	4.82

E flank

barchanoid			2.62	27.07	28.91	28.78	9.32	2.87

Crossing

crest			1.91	32.18	31.99	22.52	7.53	3.23

Site 12A
Linear

crest			6.01	45.88	30.61	15.02	1.92	0.44
mid slip			1.35	21.50	28.20	40.37	7.44	1.02
E plinth		0.85	36.77	25.47	9.05	13.95	6.67	4.44
W plinth		1.31	35.80	30.14	11.11	13.75	3.80	1.81

Appendix 3 (continued)

% in each size class (phi units)								
0.0	0.5	1.0	1.5	2.0	2.5	3.0	3.5	4.0
u west		0.05	1.63	13.21	22.27	44.60	12.11	$4.24
u east			8.26	31.99	23.58	29.87	5.04	0.98
interdune		0.86	28.62	27.17	9.75	12.58	5.70	7.29

Site 13
Linear

crest			0.49	8.11	27.46	57.94	5.39	0.80
mid slip			0.47	11.89	27.82	53.27	6.04	0.50
E plinth	0.73	9.32	31.00	20.37	7.85	11.98	7.58	6.72
W plinth	0.13	5.14	38.16	22.89	6.96	8.92	6.73	6.86
u west			0.37	5.18	36.23	51.46	5.88	0.87
u east		0.12	2.82	7.35	17.94	57.42	12.20	3.23
interdune	3.26	22.40	27.29	13.09	7.52	11.75	5.97	
E flank								
barchanoid			4.61	33.53	33.01	23.51	4.34	1.34

Site 14
Star

crest			0.26	24.44	49.04	24.89	1.24	0.10
mid slip			0.27	12.55	43.54	41.06	2.53	0.20
E plinth		0.34	20.57	48.76	14.94	12.29	2.07	0.68
u west			0.84	14.17	31.82	47.64	4.99	0.43
u east			2.03	33.84	35.15	26.03	2.26	0.51
interdune	12.79	2.32	2.36	19.59	25.05	11.96	17.79	4.55	2.06
E flank								
barchanoid			11.78	60.11	19.04	7.61	1.15	0.25
Crossing								
crest		0.04	2.84	29.53	40.53	25.04	2.53	0.49

Site 15
Linear

crest		3.34	15.74	77.94	26.36	23.16	4.21	1.34
mid slip		0.18	1.25	13.01	34.98	45.63	4.62	0.44
E plinth	1.01	11.44	33.29	19.10	8.04	11.98	7.83	5.93
W plinth	1.69	12.50	24.77	21.99	20.16	28.27	9.52	6.43
interdune	1.67	12.52	25.53	15.32	10.36	18.80	9.61	4.90
Crossing								
crest			8.21	30.11	26.99	25.58	5.49	2.09

Site 16
Zibar

crest	0.52	7.62	33.65	18.68	7.87	13.02	7.54	7.16
interdune	0.75	13.39	35.60	13.49	7.87	14.10	6.35	5.46
mid stoss	1.52	15.67	29.46	14.74	9.39	16.33	6.89	4.08

Site 16A
Linear

crest			0.37	31.87	46.63	16.44	3.36	1.16
E plinth	7.46	12.62	16.91	18.89	14.05	10.73	7.36	9.14

% in each size class (phi units)	0.0	0.5	1.0	1.5	2.0	2.5	3.0	3.5	4.0
W plinth		8.09	21.59	16.86	8.51	7.10	9.19	10.53	13.12
interdune		4.69	16.18	17.83	11.72	8.64	10.44	12.18	14.22
Site 17									
Linear									
crest				1.67	12.02	36.31	43.52	5.23	1.65
mid slip				0.29	7.73	32.71	50.82	7.04	1.32
E plinth		4.25	17.34	20.25	14.32	10.75	15.04	8.19	7.68
W plinth		0.87	5.60	15.96	20.04	20.34	25.69	6.95	3.78
interdune	0.97	10.83	14.16	12.73	10.86	14.61	24.09	7.95	5.07
Site 18									
Linear									
crest			0.04	5.68	25.82	34.59	24.68	7.26	2.21
mid slip				1.64	25.73	47.95	22.11	2.34	0.22
E plinth			2.98	21.47	31.19	24.09	10.74	4.83	2.97
W plinth			1.87	14.54	29.38	32.09	12.68	5.10	2.97
u west			0.16	16.49	38.73	23.03	12.83	5.88	2.23
interdune			3.52	13.39	22.32	27.52	17.60	7.68	5.20
E flank									
barchanoid				0.94	8.27	77.76	13.01	0.61	0.19
Site 19									
Star									
crest					0.83	38.69	54.93	5.08	0.43
mid slip					1.33	41.00	52.92	4.49	0.24
base slip					3.98	63.98	30.66	1.21	0.12
E plinth				3.46	18.89	45.80	21.40	6.76	2.72
W plinth				2.56	25.86	38.06	22.16	7.26	3.08
u west					5.51	39.08	47.75	6.49	0.81
u east				0.36	8.90	49.36	34.47	5.23	1.43
interdune			2.95	17.80	23.82	19.46	18.66	8.70	5.36
E flank									
barchanoid				1.45	17.05	42.42	34.54	5.75	2.48
Site 20									
Crescentic									
crest			0.15	3.59	45.63	33.00	11.73	4.92	0.90
Barchan									
crest			0.45	9.08	29.37	32.77	19.28	7.62	2.02
Sand sheet			0.40	14.72	32.43	12.35	4.28	11.51	14.66
Site 21									
Crescentic									
crest		0.94	6.07	26.19	23.25	15.01	13.62	11.28	3.65
mid slip		0.17	1.71	9.24	14.22	21.21	31.28	19.49	2.61
mid stoss		0.83	7.98	24.50	18.17	14.24	14.65	13.52	4.99
Zibar		3.10	18.83	22.62	12.53	8.47	10.57	13.31	7.56

% in each size class (phi units)

	0.0	0.5	1.0	1.5	2.0	2.5	3.0	3.5	4.0
Site 21A									
Crescentic									
crest		1.74	8.49	24.38	27.43	14.76	11.06	10.60	4.19
Site 22A									
Crescentic									
crest	4.97	6.06	12.26	25.50	23.00	24.00	16.14	2.73	
mid stoss	1.77	11.32	24.73	14.03	7.32	11.55	21.80	7.06	
Site 22B									
Crescentic									
crest			0.76	11.85	15.09	17.00	40.63	15.75	2.37
Site 23									
Crescentic									
crest			2.01	8.48	21.87	32.76	29.62	4.81	0.64
Barchan									
crest				2.81	20.56	40.49	31.51	4.03	0.51
mid slip				0.08	4.72	34.60	53.25	6.86	0.51
mid stoss			1.29	8.05	11.21	26.66	43.68	8.26	0.69
Site 24									
Star									
crest				0.57	18.17	49.76	22.59	2.96	0.55
W plinth		1.01	5.02	22.10	25.82	23.04	16.40	5.21	1.27
u west				16.00	50.21	16.83	11.67	4.25	0.87
arm crest				16.22	27.58	40.48	20.49	2.88	0.62
Sand sheet		4.19	13.63	18.90	18.55	13.26	10.99	10.02	7.77
Site 25									
Linear									
crest			2.56	10.58	32.48	31.13	13.56	8.29	2.64
E plinth		2.67	10.11	24.10	20.54	12.11	9.39	13.50	6.76
W plinth		1.89	10.97	23.35	17.83	11.39	9.78	15.34	8.68
interdune		3.58	14.73	22.62	17.45	11.27	9.23	12.57	6.76
Crossing									
crest				0.75	18.69	41.38	25.81	11.01	2.24
Site 26									
Linear									
crest				2.18	42.88	38.11	13.59	2.36	0.66
E plinth			0.26	10.01	38.66	27.66	15.95	4.59	1.94
W plinth			1.26	21.57	32.68	19.81	14.57	5.72	2.91
u west			4.07	31.64	25.31	14.43	14.08	6.17	3.15
interdune		0.48	2.25	27.80	30.80	16.50	12.62	5.54	2.91
E flank									
barchanoid				11.29	43.49	23.75	15.45	4.08	1.39

INDEX

Aeolian deposition 4, 133-134, 135-136, 140, 142

Algodones Dunes 27, 68, 69, 71, 111

Avalanche faces 25, 31, 39, 57, 59, 61, 64, 69, 73-74, 75, 78-80, 99, 101, 106

Barchans 8, 20, 23, 45, 109, 141, 142
    Migration rate 94-95
    Sediments 54-58, 68

Bedform climbing 5, 134, 136

Bedform hierarchies 34, 38, 124, 130-131

Benguela Current 14, 15, 18

'Berg' winds 15-18, 86-87, 96-97, 118

Calcrete 12, 13, 14, 19

Cenozoic climatic history 18-19, 132, 142, 145, 147, 149

Circulation patterns 14, 15-18, 145

Conception Bay 12, 20, 23, 43, 45, 77, 135

Crescentic dunes
    Morphology 8, 20-28, 43, 44, 45, 113-114, 126-127
    Sediments
        Grain size and sorting 54-58, 68, 73
        Internal structures 78
    Winds over 88-90

Dune alignments 44-45, 114, 116-118, 119

Dune height 26, 27, 28, 35, 39, 42-45

Dune height/spacing relations 26, 28, 35, 36, 39, 40, 120-123, 127

Dune migration rates 94-96

Dune spacing 27, 28, 35, 39, 42-44, 123-127

Dynamic equilibrium 3, 5, 110

Elizabeth Bay 8, 20, 112, 139, 148

Erosion and deposition patterns 96-110

Fog 15

Gobabeb 10, 14, 78, 145, 148

Grain size and sorting 73-74, 111-112, 123-125, Appendix 2, 3
    Comparisons between dunes 68-72
    Crescentic dunes 55-58
    In Namib Sand Sea 75-78, 139
    Linear dunes 58-63
    Star dunes 63-66
    Zibar 68-67

Gran Desierto, Mexico 26, 38, 39, 71, 78, 122

Great Escarpment 7, 8, 138

Great Sand Dunes, Colorado 58, 112, 135

Heavy minerals 52, Appendix 1

Homeb Silt Formation 14, 148

Internal sedimentary structures 78-80

Interdune areas 10, 19, 33, 39, 46
    Lacustrine deposits 132, 143-145, 148
        Sands 60, 61, 64, 66, 74-75, 77

Kalahari 47, 69, 72

Kuiseb River 5, 10, 12, 13, 19, 33, 36, 43, 118, 137, 148

Linear dunes
    Erosion and deposition patterns
        Seasonal 96, 99-101
        Spatial 97, 100-102, 106, 110
    Movement 96

Morphology 28-37, 43-46
    Controls of 113, 114, 115-119, 122, 128
Sand colour 49
Sediments
    Grain size and sorting 58-63, 68-70, 74
    Internal structures 78
Wind velocity patterns 90-93, 103-106
Wind direction patterns 93, 94

Meob Bay 20, 42, 43, 45, 77

Obib Dunes 7
Orange River 7, 8, 12, 137-139

Quaternary climatic history 18-19, 116,
    132, 143-145
Quaternary sea level changes 145-148

Rainfall 14-14, 49, 144-145

Sahara 4, 6, 7, 48, 68-69, 71, 111
Sand colours 47-50
Sand cover 23, 25
Sand grain mineralogy 52, 137-139, Ap-
    pendix 1
Sand grain morphology 50-52
Sand seas 1, 2, 3, 4, 5, 81-82, 132, 135,
    136-138
Sand sheets 41, 112
    Sediments 66-68, 68, 78
Sand sources 3, 137-138
Sand supply 111, 119-121
Sand transport rates 4, 5, 81-88, 128, 133, 139-
    140
    Direction variability 83, 85, 93-94, 112-118
    Magnitude and frequency 86-87
    Seasonable variability 83-84

Spatial variability on dunes 88-93, 1-3-106,
    129-132, 135-136
Sandwich Harbour 5, 12, 23, 43, 78, 135
Sebkhas 5, 12
Shrub coppice dunes 42
Simpson-Strzelecki 47, 71-72, 122
Skeleton Coast dunefield 7, 25, 28, 58, 68,
    122
Sossus Vlei 20, 36, 43, 69, 77, 142, 145
Speed-up factors 88-93, 103-106
Sperregebiet 8, 138
Star Dunes
    Morphology 20, 23, 36-40
        Controls of 113, 118-119, 122
    Sediments
        Grain size 63-66, 68-70, 74
        Internal structures 80

Tsauchab River 10, 13
Tsondab Flats 20, 23, 68, 71
Tsondab Sandstone Formation 10, 13, 133,
    137-139, 148
Tsondab River 10, 13, 19, 33, 43, 145
Tsondab Vlei 13, 36

White Sands, New Mexico 25, 68, 69
Winds
    Regional wind regime 15-18, 81-88, 112-114,
        116
    Over dunes 88-93, 103-106, 129-132, 135-
        136

Zibar
    Morphology 20, 30, 33, 42, 45, 46
    Grain size 66-68, 68, 70, 73-74,
        111-112